PURE MATHEMATICS

10. COMBINATIONS. PERMUTATIONS. PROBABILITIES

ANTHONY NICOLAIDES

B.Sc. (Eng.), C. Eng. M.I.E.E.
SENIOR LECTURER

P.A.S.S. PUBLICATIONS

PRIVATE ACADEMIC & SCIENTIFIC STUDIES LTD

First Published in Great Britain 1994 by

Private Academic & Scientific Studies Limited

ISB

Prin
Har

Titl

1. ns.

2.

3.

4. ns.

5. ities.

10. COMBINATIONS. PERMUTATIONS. PROBABILITIES

CONTENTS

5. PROBABILITY

6. THE MATHEMATICS OF GAMBLING

7. MISCELLANEOUS EXERCISES

8. INDEX

PREFACE

This book, which is part of the GCE A level series in Pure Mathematics covers the specialised topic of Combinations, Permutations, Probabilities.

The GCE A level series in Pure Mathematics is comprised of ten books, covering the syllabuses of most examining boards. The books are designed to assist the student wishing to master the subject of Pure Mathematics. The series is easy to follow with minimum help. It can be easily adopted by a student who wishes to study it in the comforts of his home at his pace without having to attend classes formally; it is ideal for the working person who wishes to enhance his knowledge and qualification. Combinations, Permutations, Probabilities books, like all the books in the series, is divided into two parts. In Part I, the theory is comprehensively dealt with, together with many worked examples and exercises. A step by step approach is adopted in all the worked examples. Part II of the book, a special and unique feature acts as a problem solver for all the exercises set at the end of each chapter in Part I.

I am grateful to Mr. Myat Thaw Kaung, an excellent ex-student of mine, who typeset the manuscript superbly with great care on a desktop publishing system.

I am also grateful to Mr. David Judge for contributing and thoroughly checking this book.

I am also grateful to Mr. Alex Yau for thoroughly checking this book.

The GCE A is equivalent to the new GNVQ level III.

Thank is due to the following examining body who has kindly allowed me to use questions from their past examination papers.

The University of London School Examinations. UL

A. Nicolaides

10. COMBINATIONS. PERMUTATIONS. PROBABILITIES

PART I

1. COMBINATIONS

Let us choose the first three letters from the alphabet A, B and C. In how many ways can we write these letters? *ABC, BAC, CAB, CBA, ACB, BCA*.

What is the number of combinations? There is only one combination of the three letters no matter what order they are written as above.

WORKED EXAMPLE 1

Find the number of combinations of 3 different items, taken 2 at a time. You may use the example of the first three letters of the alphabet A, B and C.

SOLUTION 1

AB, BC and *AC*, there are three combinations, of 3 letters, taken 2 letters at a time.

From the above simple examples, we can deduce a simple formula, so that we can work out readily the answer of combinations?

The number of **COMBINATIONS** from a set of n different items, taken r at a time, is $\boxed{{}^nC_r = \frac{n!}{(n-r)!\,r!}}$ where $r \leq n$.

FACTORIAL NOTATION

$n! = n$ factorial

$= 1 \times 2 \times 3 \times 4 \times .... \times (n-1) \times n$ or

$= n(n-1)(n-2) .... \times 4 \times 3 \times 2 \times 1$

$5! = 5$ factorial

$= 1 \times 2 \times 3 \times 4 \times 5 = 120$ or

$= 5 \times 4 \times 3 \times 2 \times 1.$

Alternatively the factorial natation could be used

$P_5 = 5! = 120.$

WORKED EXAMPLE 2

Find the number of combinations of 3 different letters A, B and C,

(i) taken three at a time

(ii) taken two at a time and

(iii) taken one at a time. Use the formula $^nC_r = \dfrac{n!}{(n - r)!\ r!}$

SOLUTION 2

(i) $^3C_3 = \dfrac{3!}{(3 - 3)!\ 3!} = 1$

$3! = 6;\ (3 - 3)! = 0! = 1.$

(ii) $^3C_2 = \dfrac{3!}{(3 - 2)!\ 2!} = \dfrac{1 \times 2 \times 3}{1!\ (1 \times 2)} = 3.$

(iii) $^3C_1 = \dfrac{3!}{(3 - 1)!\ 1!} = \dfrac{3!}{2!\ 1!} = \dfrac{1 \times 2 \times 3}{1 \times 2} = 3.$

BINOMIAL COEFFICIENT ^nC_r, $\binom{n}{r}$

^nC_r is called a binomial coefficient and may be denoted otherwise $\binom{n}{r}$.

So $\binom{n}{r}$ is called a binomial coefficient and ^nC_r is called the number of Combinations of n things taken r at time, the order of selection is disregarded ($r \leq n$).

USING FACTORIALS

Make a table of factorials from 0 to 9 and observe their rapid increase.

n	0	1	2	3	4	5	6	7	8	9
$n!$	0!	1!	2!	3!	4!	5!	6!	7!	8!	9!
$n!$	1	1	2	6	24	120	720	5040	40320	362880

WORKED EXAMPLE 3

There are 56 matches of football, determine the number of combinations required so that 8 matches are drawn.

SOLUTION 3

$$^{56}C_8 = \frac{56!}{(56-8)!\,8!} = \frac{56!}{48!\,8!}.$$

We can divide factorials by cancelling down

$$\frac{56!}{48!\,8!} = \frac{1\times 2\times 3\times \ldots \times 48\times 49\times 50\times 51\times 52\times 53\times 54\times 55\times 56}{(1\times 2\times 3\times \ldots \times 48)\,(1\times 2\times 3\times 4\times 5\times 6\times 7\times 8)}.$$

Using the calculator to compute factorials is obviously easier.

$$\frac{56!}{48!\,8!} = 1420494075.$$

It can be seen that in order to forecast correctly eight matches to be drawn out of 56 matches is very high, 1420494075. It can be deduced from the above that

$$\frac{n!}{(n-r)!\,r!} = \frac{n(n-1)\,(n-2)\ldots(n-r+1)}{r!}$$

$$\frac{56!}{48!\,8!} = \frac{56\times 55\times 54\times \ldots \times (56-8+1)}{8!}$$

$$= \frac{56\times 55\times 54\times 53\times 52\times 51\times 50\times 49}{1\times 2\times 3\times 4\times 5\times 6\times 7\times 8}$$

$$= 1420494075$$

where $n = 56, r = 8, n - r + 1 = 56 - 8 + 1 = 49$

$$\frac{n(n-1)(n-2)\ldots(n-r+1)}{r!} = \frac{n!}{(n-r)!\,r!}$$

Since

$$(1+x)^n = 1 + {}^nC_1\,x + {}^nC_2\,x^2 + {}^nC_3\,x^3 + \ldots + {}^nC_n\,x^n$$

$$(1+x)^n = 1 + \binom{n}{1}x + \binom{n}{2}x^2 + \binom{n}{3}x^3 + \ldots + \binom{n}{n}x^n$$

$${}^nC_1 = \frac{n!}{(n-1)!\,1!} = \frac{(n-1)!\,n}{(n-1)!\,1!} = n = \binom{n}{1}$$

$${}^nC_2 = \frac{n!}{(n-2)!\,2!} = \frac{(n-2)!\,(n-1)\,n}{(n-2)!\,2!} = \frac{(n-1)n}{2!} = \binom{n}{2}$$

$${}^nC_3 = \frac{n!}{(n-3)!\,3!} = \frac{(n-3)!\,(n-2)\,(n-1)\,n}{(n-3)!\,3!} = \frac{(n-2)\,(n-1)\,n}{3!} = \binom{n}{3}$$

∘

∘

∘

$${}^nC_r = \frac{n!}{(n-r)!\,r!} = \frac{(n-r)!\,(n-r+1)\ldots(n-2)\,(n-1)\,n}{(n-r)!\,r!}$$

$$= \frac{(n-r+1)\ldots(n-2)\,(n-1)\,n}{r!} = \binom{n}{r}$$

$${}^nC_n = \frac{n!}{(n-n)!\,n!} = \frac{1}{0!} = \frac{1}{1} = 1 = \binom{n}{n}.$$

PROPERTIES OF COMBINATIONS

To show that

$$\binom{n}{r} = \binom{n}{n-r}$$

$$\binom{n}{n-r} = \frac{n!}{[n-(n-r)]!\,(n-r)!} = \frac{n!}{(n-r)!\,r!}$$

$$\binom{n}{r} = \frac{n!}{(n-r)!\, r!}$$

$$\boxed{\binom{n}{r} = \binom{n}{n-r}} \quad \ldots (1)$$

To show that

$$\binom{n}{r} = \binom{n-1}{r} + \binom{n-1}{r-1}$$

$$\binom{n-1}{r} + \binom{n-1}{r-1} = \frac{(n-1)!}{(n-1-r)!\, r!} + \frac{(n-1)!}{[n-1-(r-1)]!\, (r-1)!}$$

$$= \frac{(n-1)!\, n}{n(n-1-r)!\, r!} + \frac{(n-1)!\, nr}{n(n-r)!\, (r-1)!\, r}$$

$$= \frac{n!}{n(n-1-r)!\, r!} + \frac{n!\, r}{n(n-r)!\, r!}$$

$$= \frac{n!}{r!}\left[\frac{1}{n(n-1-r)!} + \frac{r}{n(n-r)!}\right]$$

$$= \frac{n!}{r!}\left[\frac{n-r}{n(n-r)\,(n-1-r)!} + \frac{r}{n(n-r)!}\right]$$

$$= \frac{n!}{r!}\left[\frac{n-r+r}{n(n-r)!}\right] = \frac{n!}{r!\,(n-r)!} = \binom{n}{r}$$

$$\boxed{\binom{n}{r} = \binom{n-1}{r} + \binom{n-1}{r-1}} \quad \ldots (2)$$

If n is replaced by $n - 1$ in equation (2), we have

$$\binom{n-1}{r} = \binom{n-2}{r} + \binom{n-2}{r-1} \quad \dots (3)$$

If n is replaced by $n - 2$ in equation (2), we have

$$\binom{n-2}{r} = \binom{n-3}{r} + \binom{n-3}{r-1} \quad \dots (4)$$

Substitute (3) in (2) we have

$$\binom{n}{r} = \binom{n-2}{r} + \binom{n-2}{r-1} + \binom{n-1}{r-1} \quad \dots (5)$$

If r is replaced by $r - 1$ in equation (3), we have

$$\binom{n-1}{r-1} = \binom{n-2}{r-1} + \binom{n-2}{r-2} \quad \dots (6)$$

Substituting (6) in (5) we have

$$\binom{n}{r} = \binom{n-2}{r} + \binom{n-2}{r-1} + \binom{n-2}{r-1} + \binom{n-2}{r-2} \quad \dots (7)$$

$$\boxed{\binom{n}{r} = \binom{n-2}{r} + 2\binom{n-2}{r-1} + \binom{n-2}{r-2}}$$

EXERCISES 1

1. How many combinations do we have in the words (i) *NUMBER*
(ii) *CLOSE*

taking (a) two letters (b) three letters and (c) five letters at a time.

2. In a Greyhound race there are six dogs numbered, 1, 2, 3, 4, 5, 6. Determine the number of combinations for forecasting the first and second dog in any order.

3. In a Greyhound race there are eight dogs numbered, 1, 2, 3, 4, 5, 6, 7 and 8. Determine the number of combinations for forecasting the first three in any order.

4. Determine the following factorials:-

(i) $\frac{96!}{95!}$ (ii) $\frac{25!}{26!}$ (iii) $\frac{5!}{3!\ 2!}$ (iv) $\frac{25!}{5!\ 20!}$.

5. Determine the following:-

(i) $\binom{5}{2}$ (ii) $\binom{7}{3}$ (iii) $\binom{y}{x}$ (iv) $\binom{n}{r}$.

6. Determine the following:-

(i) ${}^{8}C_{3}$ (ii) ${}^{36}C_{8}$ (iii) ${}^{n}C_{n}$ (iv) ${}^{n}C_{1}$ (v) $\binom{n}{3}$.

7. Evaluate the following:-

(i) P_5 (ii) P_3 (iii) P_r (iv) P_n (v) P_7.

8. In a certain college there are 15 administrators and 25 lecturers. How many groups of 8 can we form so that in each group it contains

(a) 5 administrators
(b) at least 6 lecturers.

9. In a certain office there are 6 administrators and 4 technical staff. How many groups of 5 can we form so that in each group it contains

   (a) 3 administrators
   (b) at least 3 technical staff.

10. Show that (i) $\binom{n}{m} = \binom{n-1}{m} + \binom{n-1}{m-1}$

   (ii) $\binom{n}{m} = \binom{n-2}{m} + 2\binom{n-2}{m-1} + \binom{n-2}{m-2}$.

11. (i) If $\binom{n}{5} = \frac{3}{4}\binom{n}{6}$, find n.

   (ii) If $\binom{n}{26} = \frac{1}{2}\binom{n}{25}$, find n.

   (iii) If $\binom{n}{7} = \frac{1}{8}\binom{n}{8}$, find n.

2. PERMUTATIONS

PERMUTATIONS OF ITEMS THAT ARE ALL DIFFERENT

The number of different permutations obtained by taking 3 different letters and arranging all 3 of therm in every possible way equals 3! = 6.

The first three letters from the alphabet may be written in six different ways.
ABC BCA CAB ACB BAC CBA that is $3! = 1 \times 2 \times 3 = 6$ permutations.

Three Greyhound dogs are numbered 1, 2, 3. They run a race in which the order that they may finish is as follows:- 123, 231, 312, 132, 213, 321; therefore to forecast the correct order for 1st, 2nd and third we need to be aware that there are six different possible outcomes.

WORKED EXAMPLE 4

In a Greyhound racing of six dogs we require to forecast the first and second dog.

SOLUTION 4

12, 21, 13, 31, 14, 41, 15, 51, 16, 61, 23, 32, 24, 42, 25, 52, 26, 62, 34, 43, 35, 53, 36, 63, 45, 54, 46, 64, 56, 65.

There are 30 different ways of forecasting in correct order the first and second dog.

The permutations of 6 dogs taken 2 at a time can be written simply as ${}^6P_2$ where

P stands for the number of different permutations and the numbers before and after the P tell how many objects we have to choose from and how many will appear in each of the permutations.

In general the number of permutations of n things taken r at a time is

$$ {}^nP_r = \frac{n!}{(n-r)!} = \frac{nP_n}{(n-r)!}. $$

Therefore

$$ {}^6P_2 = \frac{6!}{(6-2)!} = \frac{6!}{4!} = 5 \times 6 = 30 \text{ ways.} $$

${}^nP_n$ the number of permutations of n things taken all n at a time is $n!$

WORKED EXAMPLE 5

Write the number of permutations of
(i) 7 objects taken 4 at a time
(ii) 8 objects taken 5 at a time
(iii) 10 objects taken 3 at a time
and evaluate.

SOLUTION 5

(i) ${}^7P_4$ (ii) ${}^8P_5$ (iii) ${}^{10}P_3$

(i) $${}^7P_4 = \frac{7!}{(7-4)!} = \frac{7!}{3!} = 4 \times 5 \times 6 \times 7 = 840$$

(ii) $${}^8P_5 = \frac{8!}{(8-5)!} = \frac{8!}{3!} = 4 \times 5 \times 6 \times 7 \times 8 = 6720$$

(iii) $${}^{10}P_3 = \frac{10!}{(10-3)!} = 8 \times 9 \times 10 = 720.$$

WORKED EXAMPLE 6

Compare ${}^nP_n$ and ${}^nP_r$.

SOLUTION 6

$${}^nP_n = \frac{n!}{(n-n)!} = \frac{n!}{0!} = n!$$

$${}^nP_r = \frac{n!}{(n-r)!} = \frac{{}^nP_n}{(n-r)!}$$

$$\frac{{}^nP_r}{{}^nP_n} = \frac{1}{(n-r)!} \quad \text{or} \quad \frac{{}^nP_n}{{}^nP_r} = (n-r)!$$

${}^nP_n$ is larger than ${}^nP_r$ by $(n-r)!$

WORKED EXAMPLE 7

Write down the formulae for the following permutations:-

(i) ${}^{w}P_{t}$ (ii) ${}^{r}P_{n}$ (iii) ${}^{100}P_{25}$.

SOLUTION 7

(i) ${}^{w}P_{t} = \dfrac{w!}{(w - t)!}$ (ii) ${}^{r}P_{n} = \dfrac{r!}{(r - n)!}$ (iii) ${}^{100}P_{25} = \dfrac{100!}{(100 - 25)!}$

PERMUTATIONS OF ITEMS THAT ARE NOT ALL DIFFERENT

Consider the words (i) *ACCESS* and (ii) *ASSESS*. Write out all the permutations you can make from the letters of the above words.

(i) *ACCESS* there are two sets of letters which are alike $2C$ and $2S$.

ACCESS	*SSACCE*	*SSCCAE*
ACCSSE	*SSAECC*	*SSCCEA*
CCASSE	*SSACEC*	.
CCAESS	*SASCEC*	.
CACSES	*SSECCA*	.
ESSACC	*SEACCS*	.
ECCASS	*SEACSC*	.
ECACSS	*SCCSEA*	.
ECASCS	*SCCESA*	180 different words.

Let x be the number of ways that the letters of *ACCESS* can be arranged.

$x\ 2!\ 2! = 6!$

$$x = \frac{6!}{2!\ 2!} = \frac{1 \times 2 \times 3 \times 4 \times 5 \times 6}{1 \times 2 \times 1 \times 2} = 180.$$

(ii) The word *ASSESS* can be arranged in x different ways, there are four identical letters (i.e. $4S$), 4 of the 6 letters are alike.

$x4! = 6!$

$$x = \frac{6!}{4!} = \frac{4! \times 5 \times 6}{4!} = 30.$$

The letters of the word ASSESS can be permuted in 30 different ways.

ASSESS	*ESSASS*	*SEASSS*	*SESASS*	*SSSAES*
ASSSSE	*ESSSAS*	*SAESSS*	*SESSSA*	*SSSASE*
ASSSES	*ESSSSA*	*SSASSE*	*SESSAS*	*SSSESA*
AESSSS	*SASSSE*	*SSESSA*	*SSEASS*	*SSSEAS*
EASSSS	*SASSES*	*SSSSEA*	*SSAESS*	*SSESAS*
ESASSS	*SESSAS*	*SSSSAE*	*SSSEAS*	*SSASES*.

WORKED EXAMPLE 8

Find the number of ways that the following words can be permuted:-

(i) *ASS* (ii) *ADDED* (iii) *ODD* (iv) *TUTTY*

SOLUTION 8

(i) *ASS SSA SAS.*

If the letters were different then the permutation will be 3!, since two letters are identical (*SS*) then if x is the number of permutations required, then $x\ 2! = 3!$

$$x = \frac{3!}{2!} = \frac{1 \times 2 \times 3}{1 \times 2} = 3.$$

(ii) *ADDED, DEDDA, DEDAD, DADED, DADDE, DEADD, DDDAE, DDDEA, DDAED, DDEAD, DDADE, DDEDA, ADEDD, ADDDE, AEDDD, ADDDE, EADDD, EDADD, EDDAD, EDDDA.*

If the letters were different then the permutation will be 5! since three letters are identical (*DDD*) then if x is the number of permutations required then $x\ 3! = 5!$ is the required equation.

$$x = \frac{5!}{3!} = \frac{1 \times 2 \times 3 \times 4 \times 5}{1 \times 2 \times 3} = 20.$$

(iii) *ODD*

$$x\ 2! = 3!,\ x = \frac{3!}{2!} = \frac{1 \times 2 \times 3}{1 \times 2} = 3.$$

(iv) *TUTTY*

$$x\ 3! = 5!,\ x = \frac{5!}{3!} = 4 \times 5 = 20.$$

For n things, containing $'a'$ alike of one, kind, $'b'$ alike of a second kind, $'c'$ alike of a third kind and so on...; then the number of permutations of the n things taken all together is

$$\boxed{\frac{n!}{a!\ b!\ c!\ \ldots}}$$

WORKED EXAMPLE 9

How many permutations can you make from the letters of the following words:-

(i) *PERMUTATIONS* (ii) *COMBINATIONS* (iii) *PROBABILITIES*
(d) *EVEN* (v) *ASSISTANT*.

SOLUTION 9

(i) *PERMUTATIONS*

There are two T, and the total number of letter is 12.

If x is the required number of permutations then $x\ 2! = 12!$ therefore

$$x = \frac{12!}{2!} = 239500800.$$

(ii) *COMBINATIONS*

There are two (O), two (I), two (N), and the total number of letters is 12.

$$x\ 2!\ 2!\ 2! = 12!$$

$$x = \frac{12!}{2!\ 2!\ 2!} = 59875200.$$

(iii) *PROBABILITIES*

There are two (B), three (I), and the total number of letters is 13.

$$x\ 2!\ 3! = 13!$$

$$x = \frac{13!}{2!\ 3!} = 518918400.$$

If we have attempted to work out all the different ways, it would have taken us a very long time, therefore in deducing a formula and employing it is imperative.

DEDUCTION OF THE FORMULA

$$x = \frac{n!}{a!\ b!\ c!\ \ldots}$$

where n is the number of permutations obtained from a group of n things taken all together and there are 'a' alike of one kind, 'b' alike of a second kind, 'c' alike of a third kind and so on.

$x = n!$ where all the things are different $x\ (a!\ b!\ c!\ \ldots) = n!$

GROUPS

There are eight candidates to be chosen for two Mathematics Lecturing Posts, 5 candidates are to be chosen for one Post and 3 candidates are to be chosen for the other Post. In how many ways can the 8 different candidates can be chosen. In selecting 5 candidates for the one Post you are leaving 3 for the other or vice-versa.

The one combination is ${}^8C_5$ and the other ${}^8C_3$ or ${}^8C_3$ and ${}^8C_5$.

The number of ways is

$${}^8C_5 = {}^8C_3 = \frac{8!}{3!\ 5!} = \frac{8!}{5!\ 3!} = \frac{6 \times 7 \times 8}{1 \times 2 \times 3} = 56.$$

COMBINING COMBINATIONS

Consider an example for forming a committee of 4 males and 5 females from a group of 6 males and 8 females.

${}^6C_4$ is the number of choosing the males.

${}^8C_5$ is the number of choosing the females.

$${}^6C_4 = \frac{6!}{(6 - 4)!\ 4!} = \frac{6!}{2!\ 4!} = \frac{5 \times 6}{1 \times 2} = 15$$

$${}^8C_5 = \frac{8!}{(8 - 5)!\ 5!} = \frac{6 \times 7 \times 8}{1 \times 2 \times 3} = 56.$$

The number of committees to be formed is

$${}^6C_4 \times {}^8C_5 = 15 \times 56 = 840.$$

WORKED EXAMPLE 10

A numismatist has a selection of coins from different countries, 10 Greek coins, 6 British Coins, and 8 French Coins. He decides to sell half of the Greek, half of the British and half of the French coins. How many ways can he choose the coins?

SOLUTION 10

$$ {}^{10}C_5 = \frac{10!}{(10 - 5)!\ 5!} = \frac{10!}{5!\ 5!},\quad {}^{8}C_4 = \frac{8!}{(8 - 4)!\ 4!} = \frac{8!}{4!\ 4!} $$

$$ {}^{6}C_3 = \frac{6!}{(6 - 3)!\ 3!} = \frac{6!}{3!\ 3!} $$

$$ {}^{10}C_5 \times {}^{6}C_3 \times {}^{8}C_4 = \frac{10!}{5!\ 5!} \times \frac{6!}{3!\ 3!} \times \frac{8!}{4!\ 4!} $$

$$ = \frac{6 \times 7 \times 8 \times 9 \times 10}{1 \times 2 \times 3 \times 4 \times 5} \times \frac{4 \times 5 \times 6}{1 \times 2 \times 3} \times \frac{5 \times 6 \times 7 \times 8}{1 \times 2 \times 3 \times 4} $$

$$ = 7 \times 36 \times 20 \times 70 = 252 \times 20 \times 70 = 352800. $$

WORKED EXAMPLE 11

A student has 10 different Mathematics (M) books, 5 different Electronics (E) books and 7 different Telecommunication (T) books. He wishes to prepare for his final examination and he is required to select $3M$, $2E$ and $3T$ for his revision. How many different selections can he make. How many selections can he make if he can choose any 17 books.

SOLUTION 11

$$ {}^{10}C_3 \times {}^{5}C_2 \times {}^{7}C_3 = \frac{10!}{7!\ 3!} \times \frac{5!}{3!\ 2!} \times \frac{7!}{4!\ 3!} = \frac{10! \times 5!}{10368} = 42000. $$

$$ {}^{22}C_{17} = \frac{22!}{(22 - 17)!\ 17!} = \frac{22!}{5!\ 17!} = \frac{18 \times 19 \times 20 \times 21 \times 22}{1 \times 2 \times 3 \times 4 \times 5} = 26334. $$

In how many ways we can select groups of two from 6 students.
Let the students be A, B, C, D, E, F. A can group with each of the remaining five as AB, AC, AD, AE, AF now we have B, C, D, E, F left, B can group with each of the remaining four as BC, BD, BE, BF, now we have C, D, E, F left, C can

group with each of the remaining three as *CD, CE, CF,* now we have *D, E, F* left, *D* can group with each of the remaining two as *DE* and *DF*, *E* can group with the last one *F*, *EF*. Therefore we have *AB, AC, AD, AE, AF, BC, BD, BE, BF, CD, CE, CF, DE, DF* and *EF*, fifteen groups of twos.

Alternatively

One student A can select the other one of his group in $\binom{5}{1}$ ways. If B is another student who does not belong in the group A, we can select the other one of his group is $\binom{3}{1}$ ways, the remaining two consist, the third group.

Therefore $\binom{5}{1}\binom{3}{1}\binom{1}{1} = \frac{5!}{1!\,4!} \times \frac{3!}{1!\,2!} \times \frac{1!}{1!\,0!} = 5 \times 3 \times 1 = 15.$

WORKED EXAMPLE 12

In how many ways can we divide 12 students in groups of fours.

SOLUTION 12

One student A can select the other three of his group in $\binom{11}{3}$ ways. If B is another student who does not belong in the group A, we can select the other three of his group in $\binom{7}{3}$ ways, the remaining four consist the third group. Therefore the division can be done in $\binom{11}{3}\binom{7}{3} = 5775$ ways.

WORKED EXAMPLE 13

In how many ways can we divide 15 students in groups of fives.

SOLUTION 13

One student A can select the other four of his group in $\binom{14}{4}$ ways. If B is another student who does not belong in group A, we can select the other four of his group in $\binom{9}{4}$ ways, the remaining five consist the third group. Therefore the division can be done in $\binom{14}{4} \cdot \binom{9}{4} = 1001 \times 126 = 126126$ ways.

WORKED EXAMPLE 14

One class room has 7 boys and 3 girls.

(i) In how many ways can we select a representation of twos.
(ii) How many of these ways will contain one girl.
(iii) How many ways of representations can we have with at least one girl.

SOLUTION 14

(i) A representation of twos can be selected in $\binom{10}{2}$ ways $= \frac{10!}{2!\,8!} = \frac{9 \times 10}{1 \times 2} = 45.$

(ii) One girl is selected in $\binom{3}{1} = \frac{3!}{1!\ 2!} = 3$ ways although one boy who belongs in the representation is selected in $\binom{7}{1} = \frac{7!}{6!\ 1!} = 7$ ways. Therefore we can form $\binom{3}{1}\binom{7}{1} = 3 \times 7 = 21$ representations that will contain one girl.

(iii) At least one girl, means 1 girl or 2 girls

$$\binom{3}{1}\binom{7}{1} + \binom{3}{2}\binom{7}{0} = \frac{3!}{1!\ 2!} \times \frac{7!}{6!\ 1!} + \frac{3!}{2!\ 1!} \times \frac{7!}{7!} = 3 \times 7 + 3 \times 1 = 24.$$

WORKED EXAMPLE 15

One classroom has 27 boys and 3 girls

(i) In how many ways can we select a representation of four people.
(ii) How many of these ways will contain one girl only.
(iii) How many ways will contain at least one girl.

SOLUTION 15

(i) A representation of four people can be selected in

$$\binom{30}{4} \text{ ways } = \frac{30!}{4!\ 26!} = \frac{27 \times 28 \times 29 \times 30}{1 \times 2 \times 3 \times 4} = 27 \times 7 \times 29 \times 5 = 27405.$$

(ii) One girl can be selected in $\binom{3}{1} = 3$ ways, although the three boys are

required to complete the representation in

$$\binom{27}{3} \text{ ways } = \frac{27!}{3!\ 24!} = \frac{25 \times 26 \times 27}{1 \times 2 \times 3} = 25 \times 13 \times 9 = 2925.$$

Therefore we can form $\binom{3}{1}\binom{27}{3} = 3 \times 2925 = 8775$ ways.

(iii) At least one girl we mean, 1 girl or 2 girls or 3 girls

$$\binom{3}{1}\binom{27}{3} + \binom{3}{2}\binom{27}{2} + \binom{3}{3}\binom{27}{1}$$

$$= \frac{3!}{1!\ 2!} \times \frac{27!}{24!\ 3!} + \frac{3!}{2!\ 1!} \times \frac{27!}{25!\ 2!} + \frac{3!}{0!\ 3!} \times \frac{27!}{26!\ 1!} = 9855.$$

Show that $\sum_{r=0}^{n} 5^r \binom{n}{0} = (1 + 5)^n = 1 + 5n + \frac{5^2 n(n-1)}{2!} + \ldots + 5^n = 6^n.$

Show that $\sum_{r=0}^{n} \binom{n}{r} = \binom{n}{0} + \binom{n}{1} + \binom{n}{2} + \ldots + \binom{n}{n} = (1 + 1)^n = 2^n.$

EXERCISES 2

1. Calculate the number of five letter arrangements that can be made with the letters of the word

   BOOKS.

2. Calculate the number of 11 letter arrangements that can be made with the letters of the word

   ARRANGEMENT.

3. Calculate the number of 9 letter arrangements that can be made with the letters of the word

   PERMUTING.

4. *ABBBCCCCDE* has 10 letters, find the number of permutations.

5. Evaluate the following:-

   (i) ${}^5P_3$ (ii) ${}^8P_3$ (iii) ${}^nP_r$.

6. Find the relationship between ${}^nP_r$ and ${}^nC_r$ where C stands for combination and P stands for Permutation.

7. Find the value of r if ${}^nP_r = 40320$ and $\begin{pmatrix} n \\ r \end{pmatrix} = 56$.

8. The letters A, B, C, D, E are to be used to form three letter words, how many words can we form.

9. How many four letter words can we construct from the letters A, B, C, D, E, F, G.

10. In an international committee meeting take part 7 Americans, 5 Germans, 4 British. In how many ways can they seat adjacent to each other.

11. How many five digit numbers can be formed (the first three digits to be odd and the last two digits to be even) from the digits 1, 2, 3, 4, 5, 6, 7, 8, 9.

12. (a) How many words can we form with the letters *A, B, C, D, E, O* if we use each letter once (the words do not have to have any meaning)

(b) How many of these words are alternatively vowel-consonant or vice-versa.

13. (a) How many words can be formed from the words

(i) *CEYLON*

(ii) *FORMAT*.

(b) How many of these words begin with a consonant and end with vowel.

14. Show that

$$\sum_{r=0}^{n} 2^r \binom{n}{r} = 2^0 \binom{n}{0} + 2^1 \binom{n}{1} + 2^2 \binom{n}{2} + \ldots + 2^n \binom{n}{n} = 3^n.$$

3. SELECTIONS AND ARRANGEMENTS

SELECTIONS

Selections relate to the Combinations, referred to in Chapter 1.

WORKED EXAMPLE 16

Find the number of different selections of 4 letters from the word *ABCDEF*.

SOLUTION 16

The word *ABCDEF* has six different letters, the number of different selections of 4 letters from this word will be ${}^6C_4$, that is, 4 letters to be chosen from six different letters.

$$ {}^6C_4 = \frac{6!}{(6-4)!\ 4!} = \frac{6!}{2!\ 4!} = \frac{1 \times 2 \times 3 \times 4 \times 5 \times 6}{1 \times 2 \times 1 \times 2 \times 3 \times 4} $$

$$ \boxed{{}^6C_4 = 15}\ . $$

WORKED EXAMPLE 17

Find the number of different selections of 3 letters from the word *ABCDEFGH*.

SOLUTION 17

$$ {}^8C_3 = \frac{8!}{5!\ 3!} = \frac{6 \times 7 \times 8}{1 \times 2 \times 3} = 56. $$

$$ \boxed{{}^8C_3 = 56}\ . $$

WORKED EXAMPLE 18

Find the number of different selections of 4 letters from the word *NUMBER*.

SOLUTION 18

$$ {}^6C_4 = \frac{6!}{4!\ 2!} = \frac{5 \times 6}{1 \times 2} = 15 $$

$$ \boxed{{}^6C_4 = 15}\ . $$

WORKED EXAMPLE 19

Find the number of different selections of
(i) 6 letters from the word *EXAMINATIONS*. (ii) Repeat for 5 letters.

SOLUTION 19

The word *EXAMINATIONS* contains 9 single letters (*EXAMINTOS*) and three double letters (*AA, II, NN*).

CONSIDER NO DOUBLES

Select 6 letters (no doubles)

(i) ------ *EXAMINTOS* Number of ways $= {}^9C_6 = \dfrac{9!}{6!\ 3!} = 84$

CONSIDER ONE DOUBLE ONLY

Select one double and 4 other letters.

AA---- *EXMINTOS* Number of ways $= {}^8C_4 = \dfrac{8!}{4!\ 4!} = 70$

II---- *EXAMNTOS* Number of ways $= {}^8C_4 = \dfrac{8!}{4!\ 4!} = 70$

NN---- *EXAMITOS* Number of ways $= {}^8C_4 = \dfrac{8!}{4!\ 4!} = 70$

CONSIDER TWO DOUBLES ONLY

Select two doubles and 2 other letters

AAII-- *EXMNTOS* Number of ways $= {}^7C_2 = \dfrac{7!}{5!\ 2!} = 21$

IINN-- *EXAMTOS* Number of ways $= {}^7C_2 = \dfrac{7!}{5!\ 2!} = 21$

AANN-- *EXMITOS* Number of ways $= {}^7C_2 = \dfrac{7!}{5!\ 2!} = 21$

CONSIDER THREE DOUBLES

Select 3 doubles and no others.

AAIINN-- *EXMTOS* Number of ways $= {}^6C_0 = \dfrac{6!}{6!\ 0!} = 1$

The total number of different selections $= 84 + 70 + 70 + 70 + 21 + 21 + 21 + 1 = 358$.

(ii) For different selections of 5 letters.

CONSIDER NO DOUBLES

Select 5 letters (no doubles)

------ *EXAMINTOS* Number of ways $= {}^9C_5 = \frac{9!}{4!\ 5!} = 126$

CONSIDER ONE DOUBLE ONLY

Select one double and three other letters.

AA---- *EXMINTOS* Number of ways $= {}^8C_3 = \frac{8!}{5!\ 3!} = 56$

II---- *EXAMNTOS* Number of ways $= {}^8C_3 = \frac{8!}{5!\ 3!} = 56$

NN---- *EXAMITOS* Number of ways $= {}^8C_3 = \frac{8!}{5!\ 3!} = 56.$

CONSIDER TWO DOUBLES

Select two double and one others.

AAII-- *EXMNTOS* Number of ways $= {}^7C_1 = \frac{7!}{6!\ 1!} = 7$

IINN-- *EXAMTOS* Number of ways $= {}^7C_1 = \frac{7!}{6!\ 1!} = 7$

AANN-- *EXMITOS* Number of ways $= {}^7C_1 = \frac{7!}{6!\ 1!} = 7.$

The total number of different selections $= 126 + 56 + 56 + 56 + 7 + 7 = 315.$

WORKED EXAMPLE 20

Find the number of different selections of 4 letters from the word *STATISTICS*.

SOLUTION 20

The word *STATISTICS* contains 5 single letters, 3 double letters and 2 treble letters.

The single letters are *STAIC*.
The double letters are *SS, TT, II*.
The treble letters are *SSS, TTT*.

CONSIDER NO DOUBLES OR TREBLES

---- *STAIC* Number of ways $= {}^5C_4 = 5.$

CONSIDER ONE DOUBLE ONLY

SS--	*TAIC*	Number of ways $= {}^4C_2 = 6$
TT--	*SAIC*	Number of ways $= {}^4C_2 = 6$
II--	*STAC*	Number of ways $= {}^4C_2 = 6$.

CONSIDER ONE TREBLE ONLY

SSS-	*TAIC*	Number of ways $= {}^4C_1 = 4$
TTT-	*SAIC*	Number of ways $= {}^4C_1 = 4$.

CONSIDER TWO DOUBLES

SSTT	Number of ways = 1
SSII	Number of ways = 1
TTII	Number of ways = 1.

Total number of ways = 5 + 6 + 6 + 6 + 4 + 4 + 1 + 1 + 1 = 34.

WORKED EXAMPLE 21

Find the number of different selections of 4 letters from the word *SUMMER*.

SOLUTION 21

The word *SUMMER* contains 5 single letters (*SUMER*) and one double letter (*MM*).

CONSIDER NO DOUBLES

----	*SUMER*	Number of ways $= {}^5C_4 = 5$.

CONSIDER THE ONE DOUBLE ONLY

MM--	*SUER*	Number of ways $= {}^4C_2 = 6$

The total number of different selections = 5 + 6 = 11.

WORKED EXAMPLE 22

Find the number of different selections of 4 letters from the word *NUTTER*.

SOLUTION 22

The word *NUTTER* contains 5 single letters (*NUTER*) and one double letter (*TT*).

CONSIDER NO DOUBLES

----	*NUTER*	Number of ways $= {}^5C_4 = 5$.

CONSIDER THE ONE DOUBLE ONLY

TT--	*NUER*	Number of ways $= {}^4C_2 = 6$

The total number of different selections = 5 + 6 = 11.

WORKED EXAMPLE 23

Find the number of different selections of 3 letters from the words:-
(a) *SELECTIONS*
(b) *ARRANGEMENTS*.

SOLUTION 23

(a) The word *SELECTIONS* contains 8 single letters (*SELCTION*) and two double letters (*SS, EE*).

CONSIDER NO DOUBLES

--- *SELCTION* Number of ways $= {}^8C_3 = \dfrac{8!}{5!\ 3!} = 56.$

CONSIDER THE ONE DOUBLE ONLY

SS-- *ELCTION* Number of ways $= {}^7C_1 = \dfrac{7!}{6!\ 1!} = 7$

EE-- *SLCTION* Number of ways $= {}^7C_1 = \dfrac{7!}{6!\ 1!} = 7.$

The total number of different selections $= 56 + 7 + 7 = 70.$

(b) The word *ARRANGEMENTS* contains 8 single letters (*ARNGEMTS*) and four double letters (*AA, RR, NN, EE*).

CONSIDER NO DOUBLES

--- *ARNGEMTS* Number of ways $= {}^8C_3 = \dfrac{8!}{5!\ 3!} = 56.$

CONSIDER THE ONE DOUBLE ONLY

AA-- *RNGEMTS* Number of ways $= {}^7C_1 = \dfrac{7!}{6!\ 1!} = 7$

RR-- *ANGEMTS* Number of ways $= {}^7C_1 = \dfrac{7!}{6!\ 1!} = 7$

NN-- *ARGEMTS* Number of ways $= {}^7C_1 = \dfrac{7!}{6!\ 1!} = 7$

EE-- *ARNGMTS* Number of ways $= {}^7C_1 = \frac{7!}{6!\ 1!} = 7.$

The total number of different selections $= 56 + 7 + 7 + 7 + 7 = 84.$

WORKED EXAMPLE 24

Find the number of different selections of 5 letters from the words:-

(a) *SELECTIONS*

(b) *ARRANGEMENTS*.

SOLUTION 24

(a) The word *SELECTIONS* contains 8 single letters (*SELCTION*) and two double letters (*SS, EE*).

CONSIDER NO DOUBLES

--- *SELCTION* Number of ways $= {}^8C_5 = \frac{8!}{3!\ 5!} = 56.$

CONSIDER THE ONE DOUBLE ONLY

SS-- *ELCTION* Number of ways $= {}^7C_3 = \frac{7!}{4!\ 3!} = 35$

EE-- *SLCTION* Number of ways $= {}^7C_3 = \frac{7!}{4!\ 3!} = 35.$

CONSIDER TWO DOUBLES

SSEE-- *LCTION* Number of ways ${}^6C_1 = \frac{6!}{5!\ 1!} = 6.$

The total number of different selections $= 56 + 35 + 35 + 6 = 132.$

(b) The word *ARRANGEMENTS* contains 8 single letters (*ARNGEMTS*) and four double letters (*AA, RR, NN, EE*).

CONSIDER NO DOUBLES

--- *ARNGEMTS* Number of ways $= {}^8C_5 = \frac{8!}{3!\ 5!} = 56.$

CONSIDER THE ONE DOUBLE ONLY

AA-- *RNGEMTS* Number of ways $= {}^7C_3 = \frac{7!}{4!\ 3!} = 35$

RR-- *ANGEMTS* Number of ways $= {}^7C_3 = \dfrac{7!}{4!\ 3!} = 35$

NN-- *ARGEMTS* Number of ways $= {}^7C_3 = \dfrac{7!}{4!\ 3!} = 35$

EE-- *ARNGMTS* Number of ways $= {}^7C_3 = \dfrac{7!}{4!\ 3!} = 35.$

CONSIDER TWO DOUBLES

AARR-- *NGEMTS* Number of ways $= {}^6C_1 = \dfrac{6!}{5!\ 1!} = 6$

RRNN-- *AGEMTS* Number of ways $= {}^6C_1 = \dfrac{6!}{5!\ 1!} = 6$

NNEE-- *ARGMTS* Number of ways $= {}^6C_1 = \dfrac{6!}{5!\ 1!} = 6$

AANN-- *RGEMTS* Number of ways $= {}^6C_1 = \dfrac{6!}{5!\ 1!} = 6$

AAEE-- *ANGMTS* Number of ways $= {}^6C_1 = \dfrac{6!}{5!\ 1!} = 6.$

The total number of different selections $= 56 + 35 + 35 + 35 + 35 + 6 + 6 + 6 + 6 + 6 + 6 = 232.$

WORKED EXAMPLE 25

Find the number of different selections of 3, 4 and 5 letters from the word *NUMBERING*.

SOLUTION 25

The world *NUMBERING* has 8 single letters (*NUMBERIG*) and one double letter (*NN*)

(a) different selection of 3 letters.

CONSIDER NO DOUBLES

--- *NUMBERIG* Number of ways $= {}^8C_3 = \dfrac{8!}{5!\ 3!} = 56.$

CONSIDER ONE DOUBLE ONLY

NN- *UMBERIG* Number of ways $= {}^7C_1 = \dfrac{7!}{6!\ 1!} = 7.$

The total number of different selections = 56 + 7 = 63.

(b) different selection of 4 letters.

CONSIDER NO DOUBLES

---- *NUMBERIG* Number of ways $= {}^8C_4 = \dfrac{8!}{4!\ 4!} = 70.$

CONSIDER ONE DOUBLE ONLY

NN-- *UMBERIG* Number of ways $= {}^7C_2 = \dfrac{7!}{5!\ 2!} = 21.$

The total number of different selections = 70 + 21 = 91.

(c) different selection of 5 letters.

CONSIDER NO DOUBLES

----- *NUMBERIG* Number of ways $= {}^8C_5 = \dfrac{8!}{3!\ 5!} = 56.$

CONSIDER ONE DOUBLE ONLY

NN--- *UMBERIG* Number of ways $= {}^7C_3 = \dfrac{7!}{4!\ 3!} = 35.$

The total number of different selections = 56 + 35 = 91.

ARRANGEMENTS

Arrangements relate to the permutations referred to in Chaper 2

WORKED EXAMPLE 26

Find the number of different arrangements of the word *TAKEN*.

SOLUTION 26

The word *TAKEN* has 5 different letters therefore the number of different arrangements is $5! = 120$.

WORKED EXAMPLE 27

Find the number of different arrangements of the words:-

(i) *EXOTIC*
(ii) *SINGLE*
(iii) *DOUBLES*
(iv) *TREBLING*
(v) *SHY*.

SOLUTION 27

(i) *EXOTIC* has six different letters, therefore the number of different arrangements is $6! = 720$.

(ii) *SINGLE* has six different letters, therefore the number of different arrangements is $6! = 720$.

(iii) *DOUBLES* has seven different letters, therefore the number of different arrangements is $7! = 5040$.

(iv) *TREBLING* has eight different letters, therefore the number of different arrangements is $8! = 40320$.

(v) *SHY* has three different letters, therefore the number of different arrangements is $3! = 6$.

WORKED EXAMPLE 28

A student has 9 different Pure Mathematics books, determine the number of different ways of arranging these books on a shelf.

SOLUTION 28

The number of different ways of arranging these books is $9! = 362880$.

WORKED EXAMPLE 29

Find the number of different arrangements of the following words:-

(i) *STATISTICS* (ii) *STATICS*

(iii) *MATHEMATICS* (iv) *CHEMISTRY*

(v) *THERMODYNAMICS*.

SOLUTION 29

(i) *STATISTICS*

The word has 10 letters, there are two treble letters (*TTT, SSS*) and one double letter (*II*). The number of different arrangements $= \dfrac{10!}{3!\ 3!\ 2!} = 50400.$

(ii) *STATICS* is a word with 7 letters, there are two double letters (*TT, SS*).

The number of different arrangements $= \dfrac{7!}{2!\ 2!} = 1260.$

(iii) *MATHEMATICS* is a word with 11 letters, there are three double letters (*MM, AA, TT*).

The number of different arrangements $= \dfrac{11!}{2!\ 2!\ 2!} = 4989600.$

(iv) *CHEMISTRY* is a word with 9 different letters.
The number of different arrangements $= 9! = 362880.$

(v) *THERMODYNAMICS* is a word with 14 letters, there is one double letter (*MM*).

The number of different arrangements $= \dfrac{14!}{2!}.$

WORKED EXAMPLE 30

Find the number of different arrangements of the word *MATHEMATICS* if *CS* are together.

SOLUTION 30

The word *MATHEMATICS* has 11 letters.

Consider (*CS*) as one term, there are 10 terms to permutate, then permutate *CS*

$$\frac{10!}{2!\ 2!\ 2!} \times 2! = 907200.$$

WORKED EXAMPLE 31

Find the number of different arrangements of the word *MINIMUM* if *NU* are together.

SOLUTION 31

The word *MINIMUM* has 7 letters.

Consider (*NU*) as one term, there are 6 terms to permutate consisting of an trebble letter (*MMM*) and one double letter (*II*), then permutate *NU*.

$$\frac{6!}{3!\ 2!}\ 2! = \frac{6!}{3!} = 4 \times 5 \times 6 = 120.$$

WORKED EXAMPLE 32

In how many different ways can 6 men and 6 women be arranged to play in 3 mixed doubles matches at tennis? **(U.L. June 1982)**

SOLUTION 32

1st pair one man with one of the 6 women. There are 6 ways of doing this.
Having made the 1st pairing, there are now 5 men and 5 women.
Therefore there are 5 ways of forming 2nd pairing and so on.
Therefore total number of pairings $= 6! = 720$.

For 3 matches 6 pairs are required and for each match 2 pairs are required.
Therefore number of ways selecting 2 from 6 $= {}^6C_2 = 15$.

Therefore total number of ways in which 6 men and 6 women can play 3 mixed doubles tennis matches $= 720 \times 15 = 10800$.

4. SETS

SET NOTATION

$\{\quad\}$	the set of
$n(A)$	the number of elements in the set A
$\{x:\quad\}$	the set of all x such that
$\in$	is an element of
$\notin$	is not an element of
$\varnothing$	the empty (null) set
$\mathscr{E}$	the universal set
$\cup$	union
$\cap$	intersection
A'	the complement of the set A
$f: A \to B$	f is a function under which each element of set A has an image in set B
$f: x \to y$	f is a function under which x is mapped into y
$\mathbb{N}$	the set of positive integers and zero, $\{0, 1, 2, 3, \ldots\}$
$\mathbb{Z}$	the set of integers, $\{0, \pm 1, \pm 2, \pm 3, \ldots\}$
$\mathbb{Z}^+$	the set of positive integers, $\{1, 2, 3, \ldots\}$
Q	the set of rational numbers
Q^+	the set of positive real numbers $\{x : x \in Q, x > 0\}$
$\mathbb{R}$	the set of real numbers
$\mathbb{R}^+$	the set of positive real numbers and excludes zero, $\{x : x \in \mathbb{R}, x > 0\}$
$\mathbb{R}_0^+$	the set of positive real numbers and zero, $\{x : x \in \mathbb{R}, x \geq 0\}$

$\mathbb{C}$	the set of complex numbers
$\subseteq$	is a subset of
$\subset$	is a proper subset of
$\supseteq$	contains as a subset
$\supset$	contains as proper subset
$\rightarrow$	is mapped into (in the context of a mapping)
$\rightarrow$	approaches, tends to (in the context of a limit)
$\equiv$	is congruent to or identical to
$\mathbb{R}^2$	the set of all ordered pairs of real numbers or points in a Cartesian plane
$\mathbb{R}^3$	the set of all ordered triples of real numbers or points in three dimensional Probability Cartesian space
A, B, C, etc.	events
$A \bigcup B$	union of the events A and B
$A \bigcap B$	intersection of the events A and B
$P(A)$	probability of the event A
$P(A')$	The probability of obtaining the complement of the event A
$P(A \mid B)$	conditional probability of the event A given that the event B has occurred.
$\begin{pmatrix} n \\ r \end{pmatrix}$	binomial coefficient.

DEFINITION OF A SET

What is a set?

A set is a collection of numbers, of letters, of objects.
The different letters are called elements or members of the set.

For example

A set of prime numbers is denoted $A = \{1, 2, 3, 5, 7, 11, 13, ...\}$

A set of even numbers is denoted $B = \{2, 4, 6, 8, 10, ...\}$

A set of odd numbers is denoted $C = \{1, 3, 5, 7, 9, ...\}$

TYPES OF SETS

There are three types of sets, a finite set, an infinite set and an empty or null set.

For example;

$A = \sum_{r=1}^{3} r^2 = \{1^2 + 2^2 + 3^2\}$ FINITE SET

$B = \sum_{r=1}^{\infty} r^3 = \{1^3 + 2^3 + 3^3 + ...\}$ INFINITE SET

$C = \quad = \{ \quad \}$ NULL OR EMPTY SET (NO ELEMENTS)

$\in$ = a member of the set $1 \in A$ where $A = \{1, 2, 3\}$
$\notin$ = not a member of the set $2 \notin B$ where $B = \{1, 3, 5, 7\}$

$n(A)$ = The number of elements in the set A and this is defined as the order of the set.

$n(A) = 3, n(B) = 4$

SUBSETS

If $A = \{a, b, c\}$ and $B = \{a, b, c, d, e\}$ then $A \subset B$, that is, A is a subset of B.
If $C = \{1, 3, 5, 7\}$ and $D = \{1, 2, 3, 4, 5, 6, 7\}$ then $C \subset D$, that is, C is a subset of D.
Every set has at least two subsets.

If $A = \{a, b, c\}$ and $B = \{\}$ then B is a subset A.

The subsets of A are $\{\}$, $\{a\}$, $\{b\}$, $\{c\}$, $\{a, b\}$, $\{a, c\}$, $\{b, c\}$, $\{a, b, c\}$, it is observed that there are eight subsets in a set of three elements, that is, $2^3 = 8$. It can be deduced that for n elements the number of subsets are 2^n.

UNIVERSAL SET

The prime numbers between 1 and 17 are 1, 2, 3, 5, 7, 11, 13, 17 then the **UNIVERSAL SET** is {1, 2, 3, 5, 7, 11, 13, 17} that is all the prime number up to and including 17 is $\mathscr{E} = \{1, 2, 3, 5, 7, 11, 13, 17\}$

COMPLEMENT OF SET A (A')

If $A = \{1, 2, 3, 5, 7, 11, 13, 17, 19\}$
and $\mathscr{E} = \{1, 2, 3, 4, 5, 6, 7, 8, 9, 10, 11, 12, 13, 14, 15, 16, 17, 18, 19\}$

the complement of A is $A' = \{4, 6, 8, 9, 10, 12, 14, 15, 16, 18\}$

EQUAL SETS

If $A = \{10, 7, 3, 2, 11\}$ and $B = \{2, 3, 7, 10, 11\}$ then $A = B$, the order in which the elements are written is not important the two sets A and B are exactly equal since they have the same elements.

EQUIVALENT SETS

Two sets are equivalent if they have the same number of elements $A = \{a, b, c\}$, $B = \{A, B, C\}$.

$n(A) = 3$, $n(B) = 3$.

WORKED EXAMPLE 33

Define the following terms:-

(i) A set (ii) Elements of a set, (iii) $\mathscr{E} = \{A, B, C, D, E, \ldots, Z\}$

(iv) $\mathscr{E} = \{a, b, c, d, e, f, \ldots, z\}$ (v) $C = \{1, 2, 3, 4, \ldots\}$

(vi) $D = \{2^2, 4^2, 6^2, \ldots\}$ (vii) $E = \{1^3, 3^3, 5^3, \ldots\}$

(viii) $F = \{1, 2, 3\}$	(ix) finite set	(x) infinite set
(xi) empty set	(xii) $\in$	(xiii) $n(A) = 15$
(xiv) $n(C) = 5$	(xv) $\notin$	(xvi) $A \subset B$
(xvii) A subset	(xviii) Universal subset	(xix) Complement of B
(xx) $A = B$	(xxi) $A \equiv B$.	

SOLUTION 33

(i) A set is a collection of numbers or letters, or objects.

(ii) The elements of a set are the members of a set; i.e. the members of a family consist of the father, the mother, two sons and three daughters $\{M, F, S_1, S_2, D_1, D_2, D_3\}$.

(iii) $\mathscr{E} = \{A, B, C, D, E, \ldots, Z\}$ all the capital letters of the alphabet, this is a universal set.

(iv) $\mathscr{E} = \{a, b, c, \ldots, z\}$ all the lower case letter of the alphabet, this is a universal set.

(v) $C = \{1, 2, 3, 4, \ldots\}$ all the positive integers from 1 to infinity, it is an infinite set.

(vi) $D = \{2^2, 4^2, 6^2, \ldots\}$ all the squares of the even numbers an infinite set.

(vii) $E = \{1^3, 3^3, 5^3, \ldots\}$ all the cubes of the odd numbers, an infinite set.

(viii) $F = \{1, 2, 3\}$ three positive elements, a finite set.

(ix) $A = \{3, 4, 9, 7\}$ four elements, a finite set.

(x) $B = \{1, 2, 3, 4, \ldots\}$ an infinite set.

(xi) $C = \{\quad\}$ an empty or null set, that is, no elements are included. The set of even numbers, $A = \{2, 4, 6, 8, 10\}$, have no odd numbers, a subset of A is a null, $\{\quad\}$.

(xii) $\in$ is member of the set, that is $3 \in \{3, 4, 5, 9\}$, $4 \in \{3, 4, 5, 9\}$

(xiii) $n(A) = 15$, the number of elements in the set is 15, that is, the order of the set is 15.

(xiv) $n(C) = 5$, the order of this set is 5, that is there are 5 elements.

(xv) $\notin$ is not a member of the set $3 \notin \{2, 4, 6, 8\}$, that is 3 is not a member of the set $\{2, 4, 6, 8\}$

(xvi) $A \subset B$, A is a subset of B, i.e., that 2^2 is the subset of $\{2^2, 3^2, 4^2, \ldots\}$

(xvii) $\{1, 3, 5\}$ is the subset of $\{1, 3, 5, \ldots\}$

$\{2, 4, 6\}$ is the subset of $\{2, 4, 6, \ldots\}$.

(xviii) $\mathscr{E}$ = universal subset $\{\alpha, \beta, \gamma, \ldots \omega\}$, the twenty four letters of the Greek alphabet.

(xix) If $A = \{1, 3, 5, \ldots\}$ and $\mathscr{E} = \{1, 2, 3, 4, 5, \ldots\}$ the complement of A is $A' = \{2, 4, 6, \ldots\}$.

If $B = \{2, 4, 6, \ldots\}$ and $\mathscr{E} = \{1, 2, 3, 4, 5, \ldots\}$ the complement of B is $B' = \{1, 3, 5, \ldots\}$.

(xx) $A = B$

If $A = \{3, 9, 12, 15, 35\}$ and $B = \{12, 9, 35, 3, 9\}$ then $A = B$ since the elements of A are the same to the elements of B.

(xxi) $A = \{a, b, c, \ldots, z\}$, $B = \{A, B, C, \ldots, Z\}$ the lower case letters of the English Alphabet has the same number of elements of the capital letters of the English Alphabet, therefore, $A \equiv B$, A is equivalent to B, that is $n(A) \equiv n(B)$.

WORKED EXAMPLE 34

Explain the following statements:

(i) $3 \in$ {prime factors of 24}.

(ii) If $A = \{1^2, 2^2, 3^2, \ldots, 25^2\}$ then $n(A) = 25$.

(iii) If $A = \{2, 4, 6, 8, \ldots, 30\}$ then $n(A) = 15$.

(iv) The subsets of $\{a, b, c, d\}$ are: $\varnothing$, $\{a\}$, $\{b\}$, $\{c\}$, $\{d\}$, $\{a, b\}$, $\{a, c\}$, $\{a, d\}$, $\{b, c\}$, $\{b, d\}$, $\{c, d\}$, $\{a, b, c\}$, $\{a, c, d\}$, $\{a, b, d\}$, $\{b, c, d\}$, $\{a, b, c, d\}$.

(v) The subsets of $\{\alpha, \beta, \gamma\}$ are: $\varnothing$, $\{\alpha\}$, $\{\beta\}$, $\{\gamma\}$, $\{\alpha, \beta\}$, $\{\alpha, \gamma\}$, $\{\beta, \gamma\}$, $\{\alpha, \beta, \gamma\}$.

(vi) The subsets of $\{1, 2, 3, 4, 5, 6, 7, 8, 9, 10\}$.

(vii) The proper subsets of $\{1, 2, 3, 4, 5, 6, 7, 8, 9, 10\}$ are $2^{10} - 1$.

(viii) A set has 7 elements. How many subsets can be formed from its elements?

SOLUTION 34

(i) $24 = 2 \times 2 \times 2 \times 3$, the prime factors of 24 are 2, 2, 2, 3, therefore 3 is a member of the prime factor of 24, $3 \in \{2, 2, 2, 2, 3\}$.

(ii) $A = \{1^2, 2^2, 3^2, \ldots, 25^2\}$ there are 25 elements therefore the number of elements of the set A is 25, that is, $n(A) = 25$.

(iii) $A = \{2, 4, 6, 8, \ldots, 30\}$, there are 15 even numbers, therefore there are 15 elements, hence the number of elements of the set A is 15, that is, $n(A) = 15$.

(iv) The subsets of $\{a, b, c, d\}$ are $2^4 = 16$, that is, there are 16 subsets.

(v) The subsets of $\{\alpha, \beta, \gamma\}$ are $2^3 = 8$.

(vi) The subsets of $\{1, 2, 3, 4, 5, 6, 7, 8, 9, 10\}$ are 2^{10}.

(vii) The propert subsets of (vi) are $2^{10} - 1$, null subset is not included.

(viii) $2^7 = 128$.

VENN DIAGRAMS

<u>THE UNIVERSAL SET</u> is represented by a rectangle.

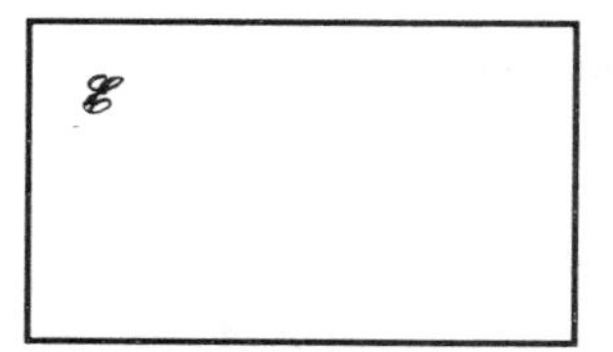

Fig. 10-I/1

<u>COMPLEMENT OF A IS A'</u>

$A \subset \mathscr{E}$

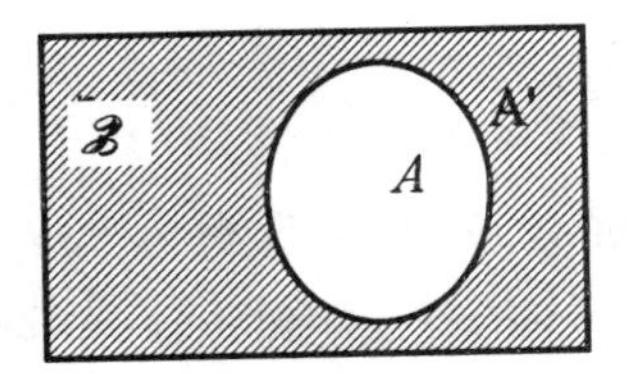

Fig. 10-I/2

$A \subset B \subset C \subset \mathscr{E}$

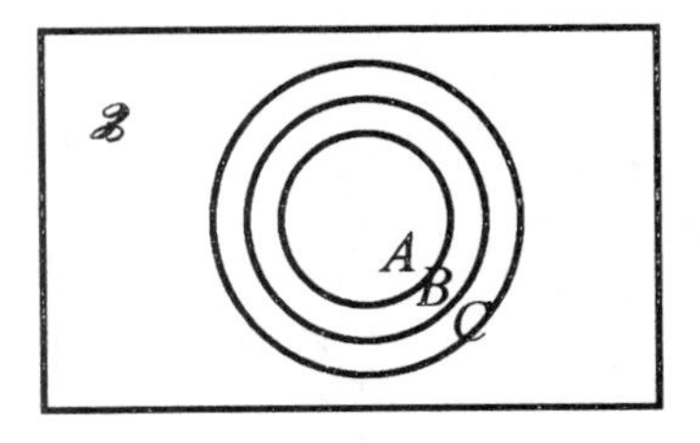

Fig. 10-I/3

UNION

$A \cup B = \{a, b, c\} \cup \{d, e, f\} = \{a, b, c, d, e, f\}$

$A \cup B$ is denoted by the shaded region

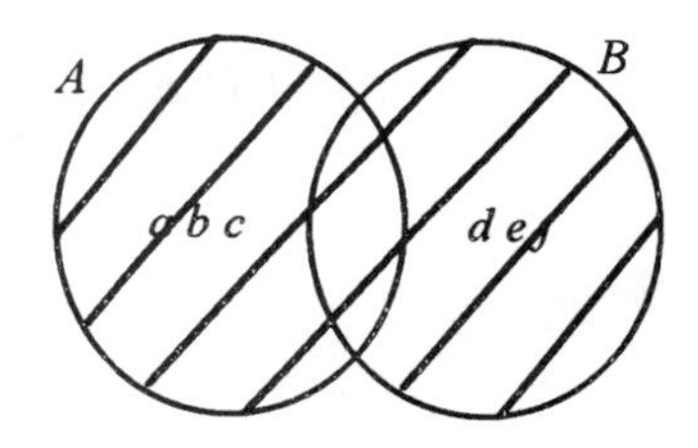

Fig. 10-I/4

$P \cup Q = \{1, 2, 3, 4\} \cup \{2, 3, 4, 5, 6\} = \{1, 2, 3, 4, 5, 6\}$

$P \cup Q$ is denoted by the shaded region

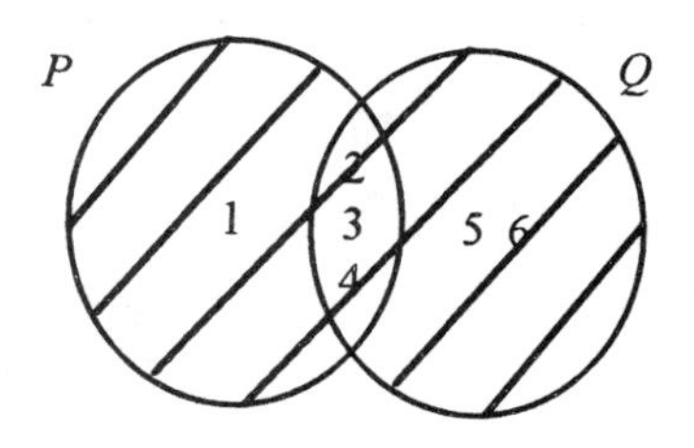

Fig. 10-I/5

INTERSECTION

$A = \{1, 3, 7, 25\}$ $B = \{2, 3, 5, 11\}$
$A \cap B = \{3\}$

$A \cap B$ is denoted by the shaded region

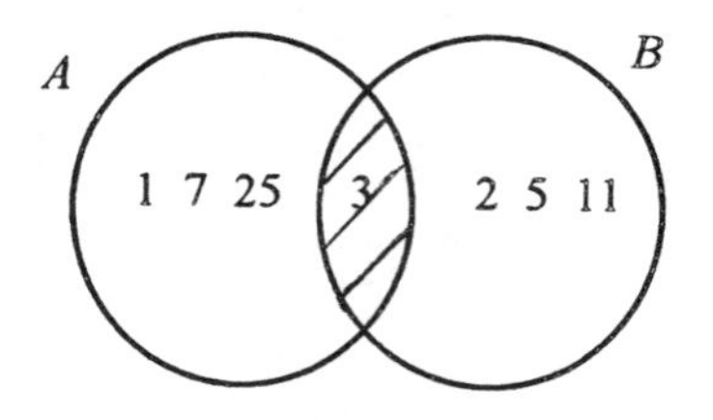

Fig. 10-I/6

WORKED EXAMPLE 35

If $A = \{1, 5, 8, 10\}$, $B = \{1, 3, 5, 7, 8, 9\}$ and $\mathscr{E} = \{0, 1, 2, 3, 4, 5, 6, 7, 8, 9, 10\}$. Draw a Venn diagram to represent these sets. Hence write down the elements of the following sets:

(i) A' (ii) B' (iii) $A \cup B$ (iv) $A \cap B$.

SOLUTION 35

The Venn diagram

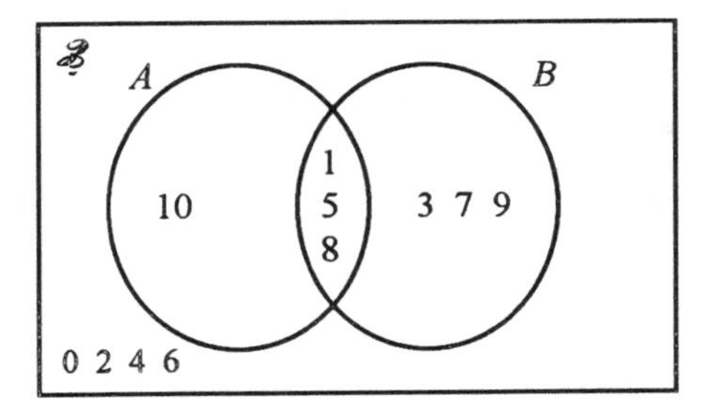

Fig. 10-I/7

(i) $A' = \{0, 2, 3, 4, 6, 7, 9\}$

(ii) $B' = \{0, 2, 4, 6, 10\}$

(iii) $A \cup B = \{1, 3, 5, 7, 8, 9, 10\}$

(iv) $A \cap B = \{1, 5, 8\}$

DISJOINT SETS

Two disjoint sets A and B, have no elements which are common to each other.

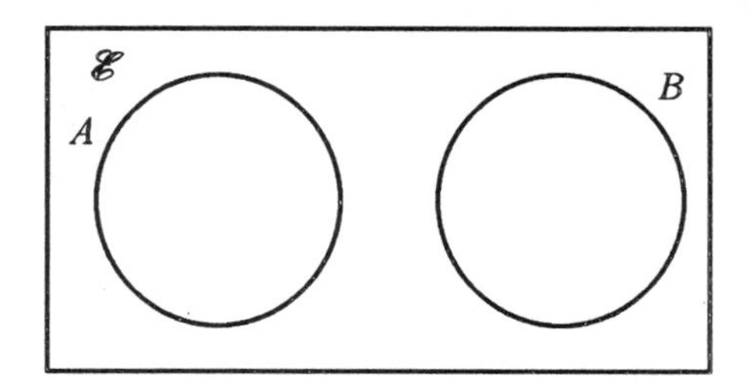

Fig. 10-I/8

INTERSECTIONS BETWEEN THREE OR MORE SETS

$A \cap B \cap C$ shows the shaded portion

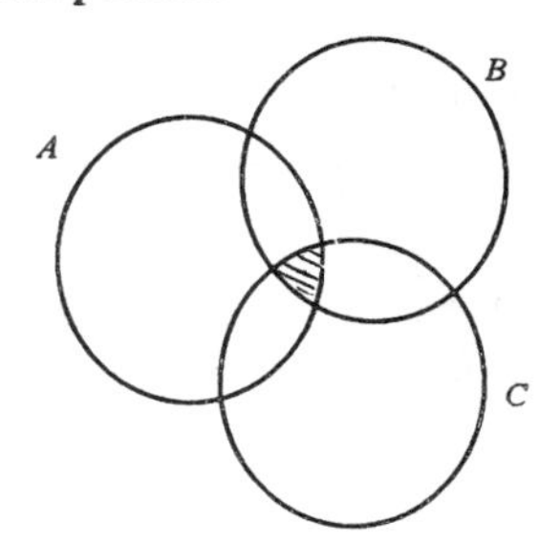

Fig. 10-I/9

WORKED EXAMPLE 36

If $A = \{1, 2, 3, 4\}$, $B = \{2, 3, 5, 6, 7\}$ and $C = \{1, 3, 5, 7, 9, 10\}$. Find

(i) $A \cup B$ (ii) $A \cup C$ (iii) $B \cup C$ (iv) $A \cap B$

(v) $A \cap C$ (vi) $B \cap C$ (vii) $(A \cap B) \cap C$

(viii) $A \cup (B \cup C)$ (ix) $(A \cap B) \cup C$ (x) $A \cup (B \cap C)$

(xi) $A \cap (B \cap C)$ (xii) $(A \cup B) \cup C$ (xiii) $A \cap (B \cup C)$.

SOLUTION 34

(i) $A \cup B = \{1, 2, 3, 4\} \cup \{2, 3, 5, 6, 7\} = \{1, 2, 3, 4, 5, 6, 7\}$

(ii) $A \cup C = \{1, 2, 3, 4\} \cup \{1, 3, 5, 7, 9, 10\} = \{1, 2, 3, 4, 5, 7, 9, 10\}$

(iii) $B \cup C = \{2, 3, 5, 6, 7\} \cup \{1, 3, 5, 7, 9, 10\} = \{1, 2, 3, 5, 6, 7, 9, 10\}$

(iv) $A \cap B = \{1, 2, 3, 4\} \cap \{2, 3, 5, 6, 7\} = \{2, 3\}$

(v) $A \cap C = \{1, 2, 3, 4\} \cap \{1, 3, 5, 7, 9, 10\} = \{1, 3\}$

(vi) $B \cap C = \{2, 3, 5, 6, 7\} \cap \{1, 3, 5, 7, 9, 10\} = \{3, 5, 7\}$

(vii) $(A \cap B) \cap C = (\{1, 2, 3, 4\} \cap \{2, 3, 5, 6, 7\}) \cap \{1, 3, 5, 7, 9, 10\} = \{3\}$

(viii) $A \cup (B \cup C) = \{1, 2, 3, 4\} \cup (\{2, 3, 5, 6, 7\} \cup \{1, 3, 5, 7, 9, 10\})$
$= \{1, 2, 3, 4\} \cup \{1, 2, 3, 5, 6, 7, 9, 10\}$
$= \{1, 2, 3, 4, 5, 6, 7, 9, 10\}$

(ix) $(A \cap B) \cup C = (\{1, 2, 3, 4\} \cap \{2, 3, 5, 6, 7\}) \cup \{1, 3, 5, 7, 9, 10\}$
$= \{2, 3\} \cup \{1, 3, 5, 7, 9, 10\} = \{1, 2, 3, 5, 7, 9, 10\}$

(x) $A \cup (B \cap C) = \{1, 2, 3, 4\} \cup (\{2, 3, 5, 6, 7\} \cap \{1, 3, 5, 7, 9, 10\})$
$= \{1, 2, 3, 4\} \cup \{3, 5, 7\} = \{1, 2, 3, 4, 5, 7\}$

(xi) $A \cap (B \cap C) = \{1, 2, 3, 4\} \cap (\{2, 3, 5, 6, 7\} \cap \{1, 3, 5, 7, 9, 10\})$
$= \{1, 2, 3, 4\} \cap \{3, 5\} = \{3\}$

(xii) $(A \cup B) \cup C = (\{1, 2, 3, 4\} \cup \{2, 3, 5, 6, 7\}) \cup \{1, 3, 5, 7, 9, 10\}$
$= \{1, 2, 3, 4, 5, 6, 7\} \cup \{1, 3, 5, 7, 9, 10\}$
$= \{1, 2, 3, 4, 5, 6, 7, 9, 10\}$

(xiii) $A \cap (B \cup C) = \{1, 2, 3, 4\} \cap (\{2, 3, 5, 6, 7\} \cup \{1, 3, 5, 7, 9, 10\})$
$= \{1, 2, 3, 4\} \cap \{1, 2, 3, 5, 6, 7, 9, 10\} = \{1, 2, 3\}$

From the above it is observed that

$$(A \cup B) \cup C = A \cup (B \cup C)$$

and,

$$(A \cap B) \cap C = A \cap (B \cap C)$$

ASSOCIATIVE LAW FOR SETS

THE NUMBER OF ELEMENTS OF A SET

TO SHOW THAT

$$n(A \cup B) = n(A) + n(B) - n(A \cap B)$$

TO SHOW THAT

$$n(A \cup B \cup C) = n(A) + n(B) + n(C) - n(A \cap B) - n(A \cap C) - n(B \cap C) + n(A \cap B \cap C)$$

Refer to the following diagram:

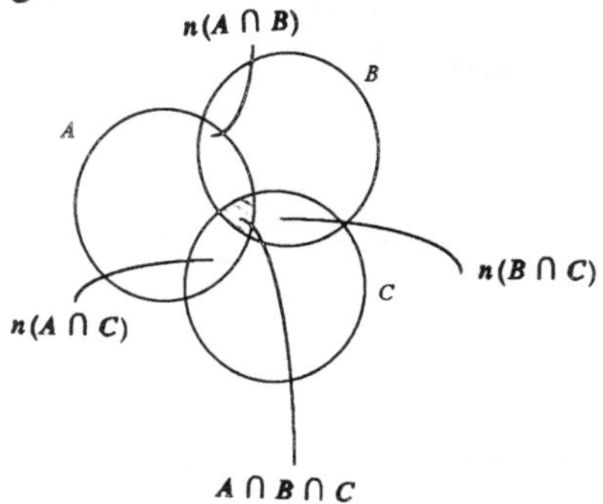

Fig. 10-I/10

WORKED EXAMPLE 37

In a class of 50 GCE A level students, 30 take Mathematics, 18 take Statistics and 10 take neither subject. How many students take both subjects?

SOLUTION 37

M = set of students take Mathematics
S = set of students take Statistics

$n(M) = 30$, $n(S) = 18$, $n(M \cup S) = 50 - 10 = 40$.

Let x be the number of students that take both Mathematics and Statistics.

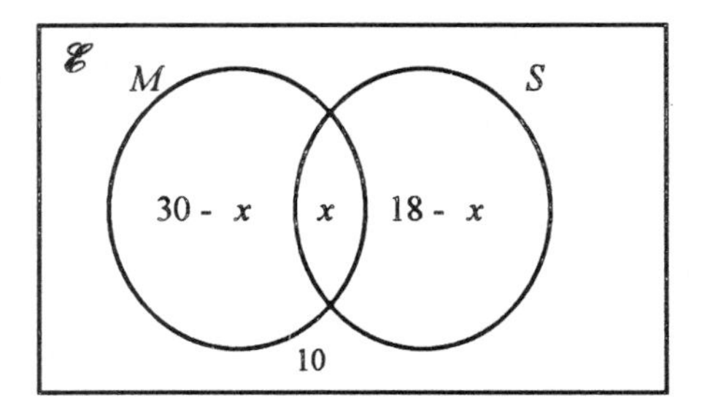

Fig. 10-I/11

$$n(M \cup S) = 30 - x + x + 18 - x = 48 - x = 50 - 10 = 40$$
$$x = 48 - 40 = 8.$$

Hence the number of students taking both subjects is 8.

Alternatively, using the equation, we have $n(M \cup S) = n(M) + n(S) - n(M \cap S)$

$n(M \cup S) = 50 - 10 = 40$
$n(M) = 30, \ n(S) = 18$

$$n(M \cup S) = n(M) + n(S) - n(M \cap S)$$
$$40 = 30 + 18 - n(M \cap S)$$
$$n(M \cap S) = 48 - 40 = 8$$

$$\boxed{n(M \cap S) = 8}$$

WORKED EXAMPLE 38

The Venn diagram of three intersections is shown in the Fig. 10-I/12 where the number of elements in the various regions are indicated.

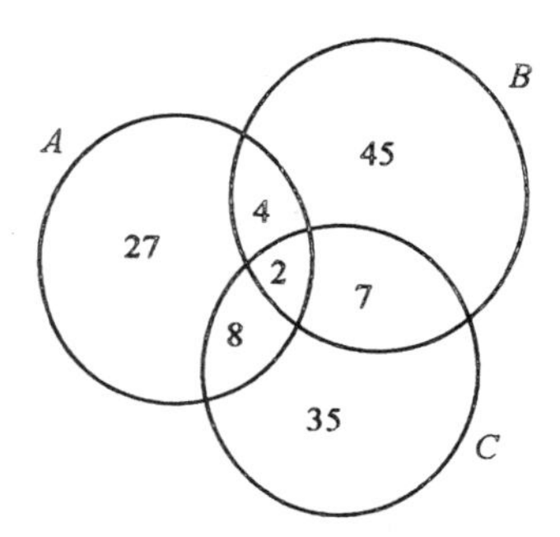

Fig. 10-I/12

Find the following:-

(i) $n(A)$ (ii) $n(B)$ (iii) $n(C)$ (iv) $n(A \cap B)$
(v) $n(A \cap C)$ (vi) $n(B \cap C)$ (vii) $n(A \cap B \cap C)$
(viii) $n[(A \cap B) \cup C]$ (ix) $n\{(A \cup B) \cup C\}$ (x) $n\{(A \cup B) \cap C\}$.

SOLUTION 38

(i) $n(A) = 27 + 4 + 2 + 8 = 41$
(ii) $n(B) = 45 + 4 + 2 + 7 = 58$
(iii) $n(C) = 35 + 8 + 2 + 7 = 52$
(iv) $n(A \cap B) = 4 + 2 = 6$
(v) $n(A \cap C) = 8 + 2 = 10$
(vi) $n(B \cap C) = 7 + 2 = 9$

(vii) $n(A \cap B \cap C) = 2$

(viii) $n[(A \cap B) \cup C] = n[(4, 2) \cup (8, 2, 7, 35)] = 56$

(ix) $n\{(A \cup B) \cup C\} = 128$

(x) $n\{(A \cup B) \cap C\} = n(93 \cap C) = 17.$

Check the formula

$$n(A \cup B \cup C) = n(A) + n(B) + n(C) - n(A \cap B) - n(A \cap C) - n(B \cap C) + n(A \cap B \cap C)$$

L.H.S. $n(A \cup B \cup C) = 128$

$n(A) = 41$, $n(B) = 58$, $n(C) = 52$, $n(A \cap B) = 6$,

$n(A \cap C) = 10$, $n(B \cap C) = 9$, $n(A \cap B \cap C) = 2$

R.H.S. $41 + 58 + 52 - 6 - 10 - 9 + 2 = 153 - 25 = 128$

WORKED EXAMPLE 39

In a particular school the numbers of candidates taking the following GCE A level examinations were as follows: 125 took Mathematics, 60 took Physics, 40 took Chemistry, 15 took all three, 25 took Mathematics and Physics, 30 took Physics and Chemistry, 20 took Mathematics and Chemistry. Determine the total number of candidates that sat these examinations.

SOLUTION 39

$$n(M \cup P \cup C) = n(M) + n(P) + n(C) - n(M \cap P) - n(P \cap C) - n(M \cap C) + n(M \cap P \cap C)$$
$$= 125 + 60 + 40 - 25 - 30 - 20 + 15 = 165$$

Alternatively

$n(M \cup P \cup C) = 95 + 5 + 15 + 10 + 20 + 15 + 5 = 165$

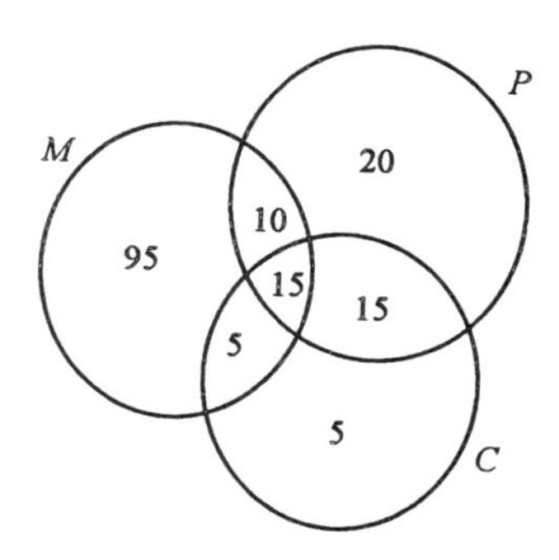

Fig. 10-I/13

5. PROBABILITY

DEFINITION

The probability of a successful outcome of an event is defined

$$P = \frac{\text{the number of ways an outcome can occur}}{\text{total number of possible outcomes}}$$

Alternatively,

PROBABILITY OF A PARTICULAR OUTCOME = $P(A)$

THE NUMBER OF EQUIPROBABLE FAVOURABLE OUTCOMES = $M(A)$

THE TOTAL NUMBER OF EQUIPROBABLE OUTCOMES = N

$$\boxed{P(A) = \frac{N(A)}{N}}$$

WORKED EXAMPLE 40

What is the probability of drawing an ace from a well-shuffled pack of cards (52). Repeat the calculation for (i) a black ace, (ii) a red ace, (iii) an ace of diamond, (iv) an ace of clubs. Recalculate the above if the pack of cards contains two jokers.

SOLUTION 40

A pack of cards consists of 52 cards: 4 aces, 4 twos, 4 threes, ..., 4 kings, 13 clubs, 13 diamonds, 13 hearts and 13 spades.

The probability of drawing an ace is $\frac{4}{52} = \frac{1}{13}$.

(i) The probability of drawing a black ace is $\frac{2}{52} = \frac{1}{26}$.

(ii) The probability of drawing a red ace is also $\frac{2}{52} = \frac{1}{26}$.

(iii) The probability of drawing an ace of diamonds is $\frac{1}{52}$.

(iv) The probability of drawing an ace of clubs is $\frac{1}{52}$

If the pack of cards contains two jokers the probabilities of drawing the above cards are altered. There are $\frac{4}{54}, \frac{2}{54}, \frac{2}{54}, \frac{1}{54}, \frac{1}{54}$ respectively.

There are now a total 54 equiprobable outcomes and the probabilities are smaller.

ZERO AND UNITY PROBABILITIES

We have observed that the probabilities lie between 0 and 1. If an event is not going to happen, the probability is 0 and if an event is certainly to take place then the probability is 1. For example if a roulette ball is spun, this will definitely land in one of the 37 numbers of a European roulette wheel, if the ball is likely to jump out of the wheel then the ball is span again, so as it will definitely will land in one of the pocket numbers, the probability is unity. Similarly if a well balanced or unbiased coin is tossed it will land either heads or tails, if for any reason it lands on its side which means it is neither a head nor a tail, it will be tossed again.

Another example, throwing a dice it will land on one of the six faces, the probability therefore of landing on one of the six faces of the cube is unity, the probability of it landing one of the corners is zero, since in physics, this will be in an astable equilibrium, like in a cone if we can try to balance the cone on its vertex.

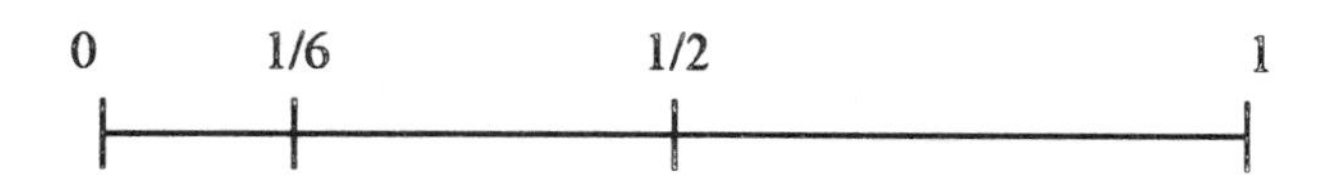

Fig. 10-I/14 SCALE OF PROBABILITY

WORKED EXAMPLE 41

In a well-shuffled pack of 54 cards (including the two Jokers), determine the following probabilities:

(i) that a card drawn at random is a club
(ii) that a card drawn at random is a red card, if a 5 or 6 or 7 of clubs are removed from the pack
(iii) that a card drawn of random is a spade, if a 2 or 3 or 4 of spades are removed from the pack.

SOLUTION 41

(i) $\frac{13}{54}$ (ii) $\frac{18}{51} = \frac{6}{17}$ (iii) $\frac{10}{51}$

For the first answer there are 13 clubs, in the second answer there are 18 reds, but the cards are reduced to 51 and finally there are only 10 spades since we have removed 3 of them.

PROBABILITY OF SUCCESS AND FAILURE

Let P and Q be the probabilities of success and failure of an even happening respectively.

$$P + Q = 1$$

Success + Failure = 1

happening + not happening = 1

The probability of obtaining a head by tossing a coin is $\frac{1}{2}$ and the probability of not obtaining a head by tossing a coin is $\frac{1}{2}$, therefore

$$\frac{1}{2} + \frac{1}{2} = 1$$

$$\boxed{P = 1 - Q} \quad \text{and} \quad \boxed{Q = 1 - P}$$

WORKED EXAMPLE 42

If the probabilities of a particular result is happening is:-

(i) $\frac{2}{3}$ (ii) $\frac{4}{5}$ (iii) $\frac{1}{6}$ (iv) 0.75 (v) 0.92.

Find the probabilities of particular result is not happening.

SOLUTION 42

(i) $\frac{1}{3}$ (ii) $\frac{1}{5}$ (iii) $\frac{5}{6}$ (iv) 0.25 (v) 0.08.

ADDITION LAWS OF PROBABILITY

MUTUALLY EXCLUSIVE EVENTS

Consider two events A and B.
The occurrence of the event A excludes the occurrence of the event B.
If the events A and B are mutually exclusive

$$P\ (A \text{ or } B) = P\ (A) + P\ (B)$$

$$\boxed{P\ (A \cup B) = P\ (A) + P\ (B)}$$ in set notation

Tossing up a coin, it can land heads or tails.
Drawing a card at random from a pack of 52 cards, the card is either red or black.
Spinning a roulette ball, the ball will land in a red or black pocket (neglecting the zero).

In the above examples, the one outcome excludes the other. The events are called mutually exclusive.

P (HEADS) + P (TAILS) = 1
P (RED) + P (BLACK) = 1

The probability of landing heads or tails is unity. The probability of drawing a red or a black card is unity.

The probability of drawing a spade, a heart or a club from a pack of 52 cards is $\frac{3}{4}$.

ADDITION LAW. VENN DIAGRAM

What is the probability of drawing a heart or an ace from a pack of 52 cards? There are thirteen hearts but one of the aces is also a heart, so there are 16 favourable outcomes, out of the 52 possible.

The probability required is $\frac{16}{52} = \frac{4}{13}$.

VENN DIAGRAM
A HEART OR AN ACE

MUTUALLY EXCLUSIVE EVENTS

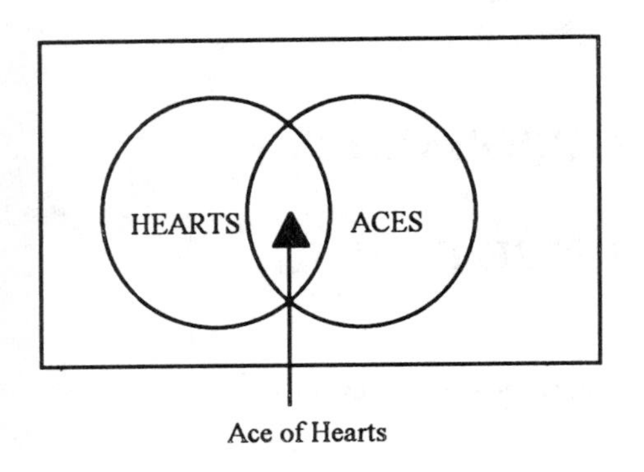

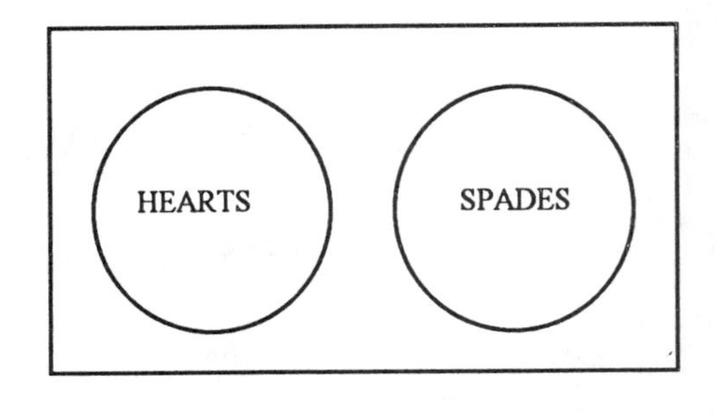

Fig. 10-I/15 ADDITION LAWS

If A and B are mutually exclusive events.

$$\boxed{P(A \text{ or } B) = P(A) + P(B)} \quad \ldots (1)$$

If A and B are not mutually exclusive events.

$$\boxed{P(A \text{ or } B \text{ or both}) = P(A) + P(B) - P(A \text{ and } B)} \quad \ldots (2)$$

WORKED EXAMPLE 43

In tossing an unbiased coin.
Find the probability of landing (i) heads (ii) tails (iii) heads or tails.

SOLUTION 43

(i) $P(A) = \dfrac{N(A)}{N} = \dfrac{1}{2}$

(ii) $P(A) = \dfrac{N(A)}{N} = \dfrac{1}{2}$

(iii) $P(A) = \dfrac{N(A)}{N} + \dfrac{N(A)}{N} = \dfrac{1}{2} + \dfrac{1}{2} = 1.$

WORKED EXAMPLE 44

A card is randomly selected from a pack of 52 playing cards. Find the probability of drawing an ace or a jack.

SOLUTION 44

$$P\text{ (Ace or Jack)} = P\text{ (Ace)} + P\text{ (Jack)}$$

$$= \frac{4}{52} + \frac{4}{52} = \frac{8}{52} = \frac{2}{13}.$$

NON MUTUALLY EXCLUSIVE EVENTS

If events A and B are not mutually exclusive

$$P\ (A \text{ or } B) = P\ (A) + P\ (B) - P\ (A)\ P\ (B)$$

$$P\ (A \cup B) = P\ (A) + P\ (B) - P\ (A \cap B)$$

WORKED EXAMPLE 45

A card is randomly drawn from a pack of 52 playing cards. Find the probability of drawing an Ace (A) or a heart (H).

SOLUTION 45

$$P\ (A \text{ or } H) = P\ (A) + P\ (H) - P\ (A)\ P\ (H)$$

$$= \frac{4}{52} + \frac{13}{52} - \frac{4}{52} \times \frac{13}{52} = \frac{17}{52} - \frac{1}{52} = \frac{16}{52} = \frac{4}{13}.$$

Observe that in a pack of cards there are 13 hearts (one of which is an ace) and 3 other aces, therefore there are 16 such cards. The total number of possible outcomes is 52 and the number of ways an outcome can occur is 16, therefore the

probability is $\frac{16}{52} = \frac{4}{13}$.

MULTIPLICATION LAW OF PROBABILITY

$$P(A_1 \text{ and } B_2) = P(A_1) \times P(B_2 \mid A_1)$$

$P(B_2 \mid A_1)$ means the probability of B_2 occurring given A_1 occurred first.

$$P(A_1 \cap B_2) = P(A_1) \times P(B_2 \mid A_1)$$

WORKED EXAMPLE 46

If a dice is thrown twice, find the probability of throwing a 'six' following by a 'five', that is, a 'six' on the first throw and a 'five' on the second throw.

SOLUTION 46

$$P(6_1 \text{ and } 5_2) = P(6_1) \times P(5_2) = \frac{1}{6} \times \frac{1}{6} = \frac{1}{36}.$$

WORKED EXAMPLE 47

Remove two cards from a pack of 52 playing cards, without replacement. Find the probability of drawing two kings (K)

SOLUTION 47

$$P(k_1 \text{ and } k_2) = P(k_1) \times P(k_2) = \frac{4}{52} \times \frac{3}{51} = \frac{1}{13} \times \frac{1}{17} = \frac{1}{221}.$$

Note that on the second draw, there are three kings left in 51 cards.

SET NOTATION ELEMENTARY EVENTS

Mutually exclusive outcomes of a random experiment.

THROWING A PAIR OF DICE UNION AND INTERSECTION OF TWO EVENTS

UNION OF TWO EVENTS A_1 and A_2 is denoted by

$$\boxed{A_1 \cup A_2}$$

A_1 UNION A_2

This may be represented by a Venn diagram

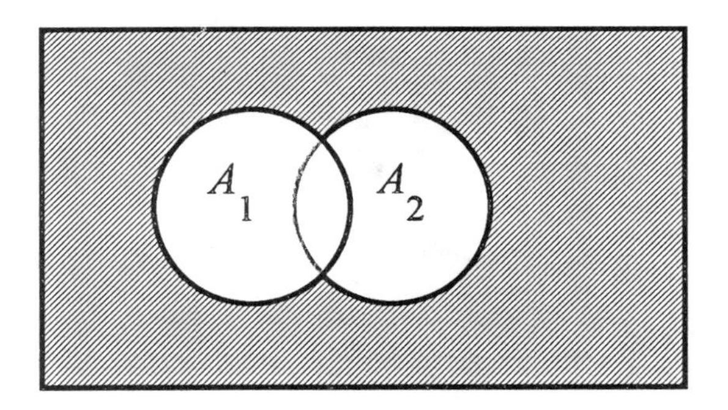

Fig. 10-I/16 UNSHADED FIGURE

or simply by

$A_1 + A_2$ BOOLEAN EXPRESSION

the occurrence of at least one of the events A_1 and A_2, that is A_1 occurs OR A_2 occurs or A_1 and A_2 both occur.

INTERSECTION OF TWO EVENTS A_1 and A_2 is denoted by

$$\boxed{A_1 \cap A_2}$$

or simply by

$A_1 A_2$ BOOLEAN EXPRESSION

the occurrence of both events A_1 AND A_2 is displayed in the unshaded area in the following Venn diagram.

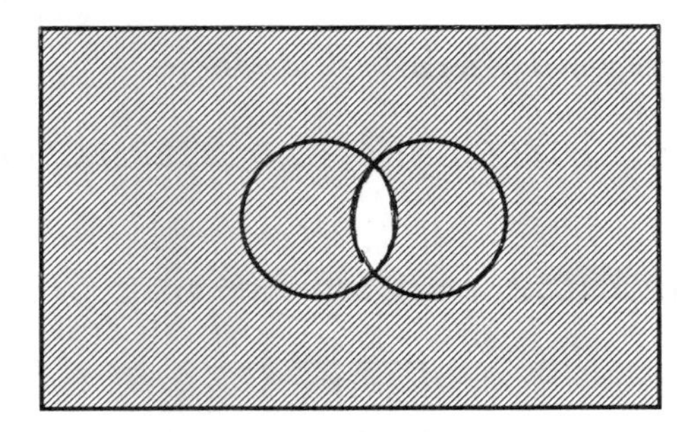

Fig. 10-I/17 UNSHADED FIGURE

Let us look at other conditions when A_1 and A_2 are mutually exclusive. This is represented by the unshaded areas in the Venn diagram below.

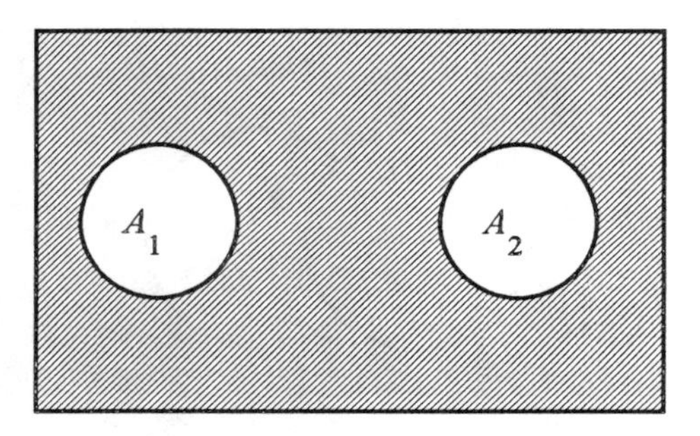

Fig. 10-I/18 UNSHADED FIGURES

The difference of $A_1 - A_2$ is represented by the unshaded areas in the Venn diagram below.

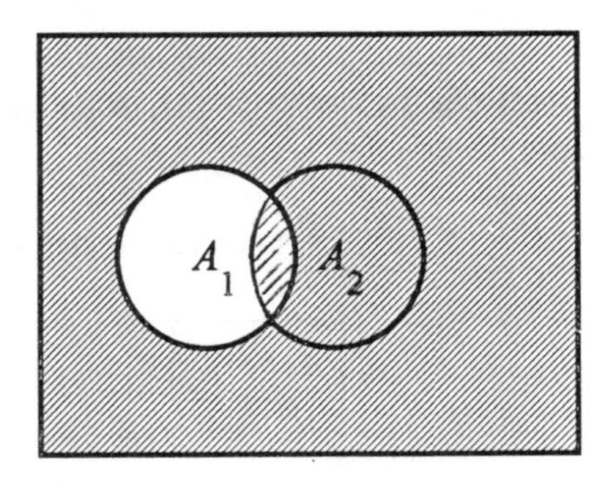

Fig. 10-I/19 UNSHADED FIGURE

A_1 occurs but not A_2

A_1 complements A_2 is represented by the unshaded area in the Venn diagram below.

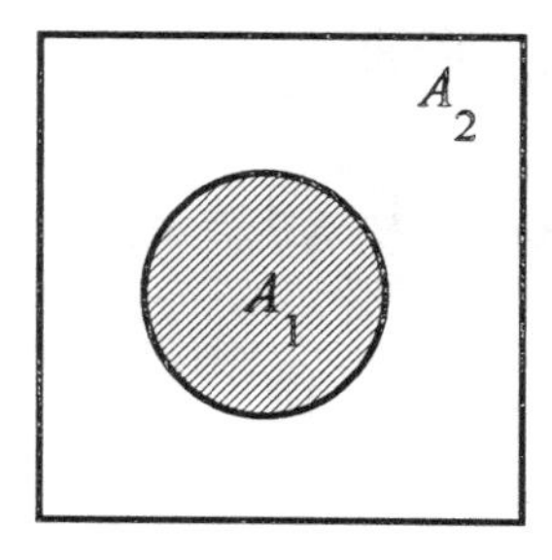

Fig. 10-I/20

The shaded and unshaded events (A_1 and A_2) are complements of each other, A_1 does not occur, A_2 occurs.

The complement of A is (NOT A) is denoted by $\bar{A}$

Subsets

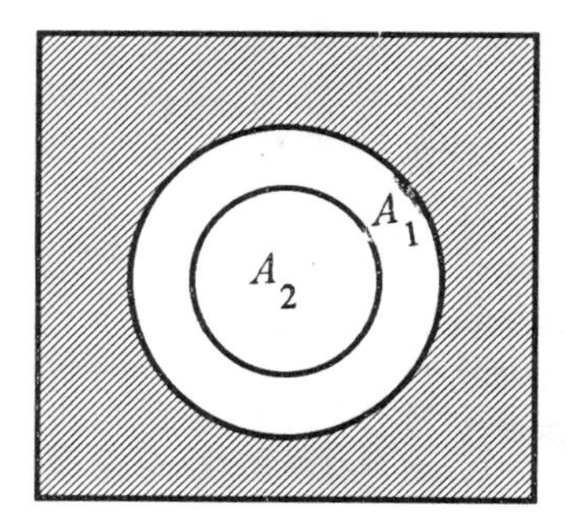

Fig. 10-I/21 $A_1 \subset A_2$ (the set A_1 is a subset of the set A_2)

$\subset$ "is a subset of" or "is contained in"
$\supset$ "contains"

In throwing a pair of dice
A_1 the event that both dice turn up even
A_2 the event that both dice turn up odd
A the event that the total number of spots is even
A_1 and A_2 are mutually exclusive.

$$A = A_1 \cup A_2$$

$$A_1 = A - A_2$$

$$A_2 = A - A_1$$

$\bar{A}$ the event that the total number of spots is odd

$\overline{A_1}$ the event that at least one die turns up odd

$\overline{A_2}$ the event that at least one die turns up even

$$\overline{A_1} - \overline{A} = \overline{A_1} - A = A_2, \quad \overline{A_2} - \overline{A} = \overline{A_2} - A = A_1$$

WORKED EXAMPLE 48

In a pack of 52 playing cards, we have:

P (HEARTS) $= \frac{1}{4}$ $\quad$ P (H or C) $= \frac{1}{2}$

P (CLUBS) $= \frac{1}{4}$ $\quad$ P (H or D) $= \frac{1}{2}$

P (DIAMONDS) $= \frac{1}{4}$ $\quad$ P (C or D) $= \frac{1}{2}$

P (SPADES) $= \frac{1}{4}$ $\quad$ P (C, D or S) $= \frac{3}{4}$

Where C represents Clubs
D represents diamonds
S represents spades
H represents hearts

Using the addition laws (1) and (2)

Determine (i) P (D or a ten)
(ii) P (C or Queen)
(iii) P (S or King)
(iv) P (H or an eight)
(v) P (S or D)

SOLUTION 48

(i) $P(D) = \frac{1}{4}$ $\quad$ P (ten) $= \frac{1}{13}$

P (ten of D) $= \frac{1}{52}$

$$\therefore\ P(D \text{ or a ten}) = P(D) + P(\text{ten}) - P(\text{ten of } D)$$

$$= \frac{1}{4} + \frac{1}{13} - \frac{1}{52} = \frac{17}{52} - \frac{1}{52} = \frac{16}{52} = \frac{4}{13}.$$

(ii) $P(C) = \frac{1}{4}$, $P(\text{Queen}) = \frac{1}{13}$, $P(\text{Queen of } C) = \frac{1}{52}$

$$P(C \text{ or Queen}) = P(C) + P(\text{Queen}) - P(\text{Queen of } C)$$

$$= \frac{1}{4} + \frac{1}{13} - \frac{1}{52} = \frac{4}{13}$$

(iii) $P(S \text{ or king}) = \frac{4}{13}$

(iv) $P(H \text{ or an eight}) = \frac{4}{13}$

(v) $P(S \text{ or } D) = P(S) + P(D) = \frac{1}{4} + \frac{1}{4} = \frac{1}{2}.$

DEPENDENT AND INDEPENDENT EVENTS

Consider a pack of 52 cards, that is, 26 red cards and 26 black cards.

The probability of drawing of random a black card is $\frac{26}{52} = \frac{1}{2}$.

If the black card is now replaced in the pack and a black card is again drawn at random, the probability of drawing a black card is again $\frac{1}{2}$. Both these events are **INDEPENDENT**.

But if the card drawn from the pack is black and is not replaced, then the probability is different, there are now 51 cards, 25 blacks and 26 reds, the probability of drawing a black card this time is $\frac{25}{51}$.

If the card drawn in the first instance was red and not replaced, the probability of drawing a black card is $\frac{26}{51}$. In the former case the probability is slightly decreased and the latter case is slightly increased. The draws are now **DEPENDENT**. Distinguish between dependent and independent events, think of some examples.

INDEPENDENT EVENTS

In the case, the outcome of one event does not affect the outcome of the other.

Consider a few examples to illustrate the above.

WORKED EXAMPLE 49

Three dice, A, B and C are thrown on to a table, what is the probability of obtaining three sixes.

SOLUTION 49

$$\boxed{P\,(A \text{ AND } B \text{ AND } C) = P(A) \times P(B) \times P(C)}$$

$$P(A) = \frac{1}{6}, \quad P(B) = \frac{1}{6}, \quad P(C) = \frac{1}{6}$$

$$P\,(A \text{ AND B AND } C) = \frac{1}{6} \times \frac{1}{6} \times \frac{1}{6} = \frac{1}{216}.$$

WORKED EXAMPLE 50

What is the probability of obtaining eight consecutive reds with eight spins of the roulette ball?

(a) Assume that there are 18 reds and 18 blacks only on the wheel.

(b) Assume that there are 18 reds, 18 blacks and one green (zero) (the stake is lost).

(c) Assume that there are 18 reds, 18 blacks and one green (zero) (half the stake is lost).

What is the percentage advantage, for the house in each of these three cases.

SOLUTION 50

(a) The probability of obtaining one red is $\frac{1}{2}$, the probability of obtaining eight consecutive reds will be

$$\frac{1}{2} \times \frac{1}{2} \times \frac{1}{2} \times \frac{1}{2} \times \frac{1}{2} \times \frac{1}{2} \times \frac{1}{2} \times \frac{1}{2} = \frac{1}{2^8} = \frac{1}{256} = 3.90625 \times 10^{-3}$$

Since $P(A \text{ AND } B \text{ AND } C \text{ AND } D \text{ AND } E \text{ AND } F \text{ AND G AND } H)$

$$= P(A.B.C.D.E.F.G.H)$$

$$= P(A) \times P(B) \times P(C) \times P(D) \times P(E) \times P(F) \times P(G) \times P(H)$$

(b) There are now 37 possible outcomes. The probability of obtaining a red is

$\frac{18}{37}$ and the probability of eight consecutive reds is similarly

$$\left(\frac{18}{37}\right)^8 = 3.137373676 \times 10^{-3}.$$

(c) If zero comes up, it is neither red nor black, and half of the stake (wager) is lost.

The total number of outcomes are 37 and chance of winning is 18.5.

The probability is $\left(\frac{18.5}{37}\right)^8 = 0.00390625$.

House advantage.

(a) zero

(b) $\frac{1}{37} = 0.027$ or $+$ 2.7%. In every 37 spins, he loses 1 piece or unit.

(c) $\frac{1/2}{37} = 0.0135$ or $+$ 1.35%.

That is, in every $2 \times 37 = 74$ spins he loses 1 piece or 1 unit.

BINOMIAL DISTRIBUTION

A Binomial distribution relates to events for which there are two possible outcomes, i.e., a pack of 52 cards is well shuffled, 26 cards are red and 26 cards are black.

A perfect roulette wheel has 36 numbers (Without the zero) has 18 reds and 18 blacks.

An unbiased coin has a head one side and a tail on the other.
Consider spinning a roulette wheel which has 36 numbers. The probability can be evaluated after working out the relevant conbination.

Spinning the roulette wheel twice

$$P\text{ (2 reds)} = \frac{1}{2} \times \frac{1}{2} = \frac{1}{4} \qquad {}^2C_2 = \frac{2!}{0!\ 2!} = 1$$

$$P\text{ (1 red)} = \frac{2}{4} = \frac{1}{2} \qquad {}^2C_1 = \frac{2!}{1!\ 1!} = 2$$

$$P\text{ (0 red)} = \frac{1}{4} = P\text{ (2 blacks)}$$

$$= \frac{1}{2} \times \frac{1}{2} = \frac{1}{4} \qquad {}^2C_0 = \frac{2!}{2!\ 0!} = 1$$

Spinning the roulette wheel three times

$$P\text{ (3 reds)} = \frac{1}{2} \times \frac{1}{2} \times \frac{1}{2} = \frac{1}{8}$$

$$P\text{ (2 reds)} = \frac{3}{8} \qquad {}^3C_2 = \frac{3!}{2!\ 1!} = 3$$

$$P\text{ (1 red)} = \frac{3}{8} \qquad {}^3C_1 = \frac{3!}{1!\ 2!} = 3$$

$$P\text{ (0 red)} = \frac{1}{8} \qquad {}^3C_0 = \frac{3!}{3!\ 0!} = 1$$

Spinning the roulette wheel four times

$$P\text{ (4 reds)} = \frac{1}{16}$$

$$P\text{ (3 reds)} = \frac{4}{16} \qquad {}^4C_3 = \frac{4!}{1!\ 3!} = 4$$

$$P\text{ (2 reds)} = \frac{6}{16} \qquad {}^4C_2 = \frac{4!}{2!\ 2!} = 6$$

$$P\text{ (1 red)} = \frac{4}{16} \qquad {}^4C_1 = \frac{4!}{4!\ 1!} = 4$$

$$P\text{ (0 red)} = \frac{1}{16} \qquad {}^4C_0 = \frac{4!}{4!\ 0!} = 1$$

PASCAL'S TRIANGLE

$$\begin{array}{ccccccccccc} & & & & 1 & & 1 & & & & \\ & & & 1 & & 2 & & 1 & & & \\ & & 1 & & 3 & & 3 & & 1 & & \\ & 1 & & 4 & & 6 & & 4 & & 1 & \\ 1 & & 5 & & 10 & & 10 & & 5 & & 1 \end{array}$$

Spinning the wheel five times (refer to the fifth line of Pascals triangle)

$$P\text{ (5 reds)} = \frac{1}{32}$$

$$P\text{ (4 reds)} = \frac{5}{32}$$

$$P\text{ (3 reds)} = \frac{10}{32}$$

$$P\text{ (2 reds)} = \frac{10}{32}$$

$$P\text{ (1 red)} = \frac{5}{32}$$

$$P\text{ (0 red)} = \frac{1}{32}.$$

To find the probability of 5 reds out of 9 spins of the wheel, then there will be $2^9 = 512$ equiprobable outcomes

$$^9C_5 = \frac{9!}{(9-5)!\,5!} = \frac{9!}{4!\,5!} = 126.$$

where $\quad ^nC_r = \dfrac{n!}{(n-r)!\,r!}.$

The probability is $\dfrac{126}{512} = \dfrac{63}{256}.$

In tossing a coin, there are two outcomes, a HEAD and a TAIL. When drawing a card from a pack of 52 playing cards, there are two outcomes, a BLACK card and a RED card. In spinning a roulette wheel, there are two outcomes, a RED number and a BLACK number.

The outcomes of the above trials are two (binomial). Let these outcome be P (Red) and Q (Black), $P(P) = p$, $P(Q) = q$. If trial is repeated several times, the outcomes are given in the tree diagram is shown below.

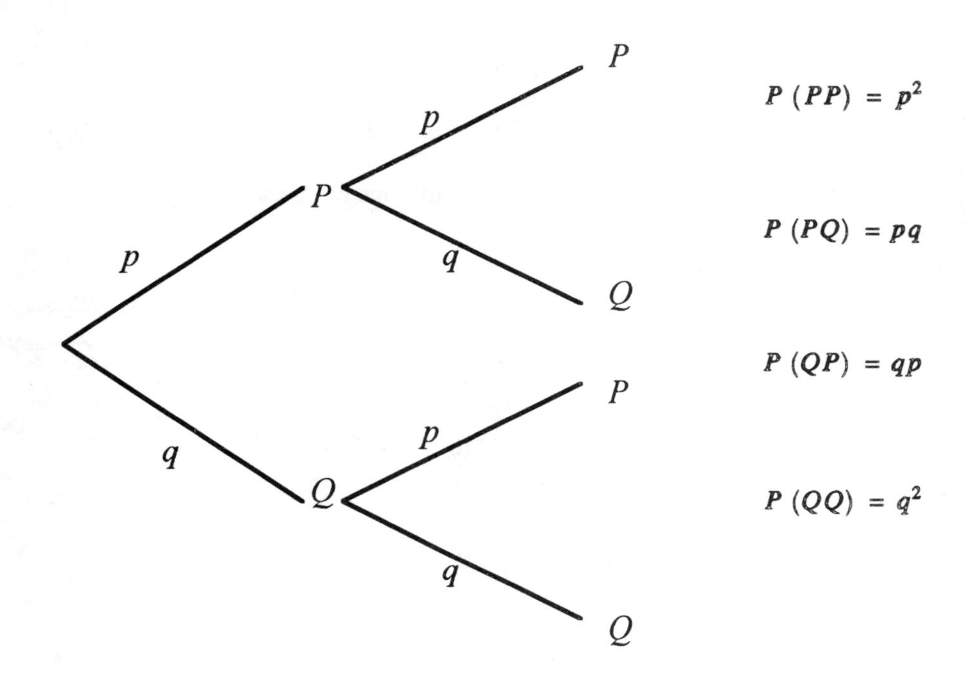

Fig. 10-I/22

$$P(PP) + P(PQ) + P(QP) + P(QQ) = p^2 + pq + qp + q^2$$
$$= p^2 + 2pq + q^2 = (p + q)^2$$

For three trials $(p + q)^3 = p^3 + 3p^2q + 3pq^2 + q^3$

For four trials $(p + q)^4 = p^4 + 4p^3q + 6p^2q^2 + 4pq^3 + q^4$

For the above coefficients, we can use the PASCAL TRIANGLE

ADDITION LAW. SET NOTATION

$A \cup B$ (A UNION B) is the set of all events favourable for A or B or both

$$\boxed{P(A \cup B) = P(A) + P(B) - P(A \cap B)}$$

where $A \cap B$ (A intersection B) is the set of all outcomes favourable for both, that is if A is the set of all diamond, B the set of all aces, then $A \cap B$ will contain just the ace of clubs.

WORKED EXAMPLE 51

A bag contains 13 balls, 7 white and 6 reds. One ball is drawn at random and not replaced. What is the probability of drawing a white ball, what is the probability that another ball drawn is also white? Find the probability of drawing two white balls.

SOLUTION 51

The probability of drawing a white ball is $\frac{7}{13}$, the ball is not replaced so that there are now 12 balls left, 6 white and 6 reds.

The probability of drawing another white ball will be $\frac{6}{12}$ or $\frac{1}{2}$. The total probability of drawing two white balls will therefore be $\frac{7}{13} \times \frac{1}{2} = \frac{7}{26}$

WORKED EXAMPLE 52

An unbiased dice is thrown five times. Find the probability that
(i) five 'threes' are shown (ii) four sixes are shown (iii) no fours are shown
(iv) at least one 'two' is shown.

SOLUTION 52

(i) P (first throw shows a three) $= \frac{1}{6}$

P (second throw shows a three) $= \frac{1}{6}$

P (all five throws shows a three) $= \left(\frac{1}{6}\right)^5$

(ii) P (all five throws shows a six) $= \left(\frac{1}{6}\right)^5$

(iii) P (none of the throws is four) $= \left(\frac{5}{6}\right)^5$

(iv) P (at least one two is shown) $= 1 - \left(\frac{5}{6}\right)^5 = \frac{6^5 - 5^5}{6^5} = \frac{4651}{7776}$.

COMPLEMENT

The complement of A is denoted by $\overline{A}$ or A', so the complement of $P(A)$ is $P(\overline{A})$ or $P(A')$

$$P(A) + P(\overline{A}) = 1$$

$$P(\overline{A}) = 1 - P(A)$$

$$P(A \cap B) = P(A) \times P(B)$$

$$P(A \cap \overline{B}) = P(A) \times P(\overline{B}) = P(A) \times [1 - P(B)] = P(A) - P(A)\,P(B)$$

$$\boxed{P(A \cap \overline{B}) = P(A) - P(A \cap B)}$$

CONDITIONAL PROBABILITY (DEPENDENT EVENTS)

The probability of A occurring under the condition that B is known to have occurred is defined by

$$\boxed{P(A \mid B) = \frac{P(A \cap B)}{P(B)}} \quad \ldots (1)$$

where $A \cap B$ is the intersection of the events A and B and it is assumed that $P(B) > 0$. A and B are dependent events.

$$0 \leq P(A \mid B) \leq 1$$

From (1) we have $P(A \cap B) = P(A/B)\,P(B)$

$$P(B \mid A) = \frac{P(A \cap B)}{P(A)} \quad \ldots (2) \text{ provided } P(A) \neq 0$$

$P(B \mid A)$ is the conditional probability B given that A has already occurred.

$$P(A \mid B) = \frac{P(A \cap B)}{P(B)} \quad \ldots (3) \text{ provided } P(B) \neq 0$$

From (2) and (3)

$$\boxed{P(A \cap B) = P(A \mid B)\,P(B) = P(A)\,P(B \mid A)}$$

If A and B are incompatible, so that $A \cap B = \emptyset$, then $P(A \mid B) = 0$

If B implies A, so that $B \subset A$, then $P(A \mid B) = 1$.

Suppose A and B are independent events then $P(B \mid A) = P(B)$

If A_1 is the event that a 'head' is obtained on the first toss of a coin and B_2 is the event that a tail is obtained on the second toss of a coin. If B_1 is the event of a tail being obtained on the first toss of a coin.

Then $P(B_2 \mid A_1) = \frac{1}{2}$ $\qquad P(B_1) = \frac{1}{2}$.

Therefore as $P(B_2 \mid A_1) = P(B_1)$, A and B are independent events.

Suppose A_1 is the event that the first card drawn without replacement from a pack of 52 playing cards, is a heart and B_2 is the event that the second card drawn is an ace. If B_1 is the event of drawing an ace on the first draw then if A_1 is the event that the ace of hearts is obtained on the first draw.

Then $P(B_2 \mid A_1) = \frac{3}{51}$ and $P(B_1) = \frac{4}{52}$

since $P(B_2 \mid A_1) \neq P(B_1)$ A and B are dependent events.

If A_1 is the event that the first card drawn is a heart (not the ace) and B_2 is the event that the second card drawn is an ace. If B_1 is the event drawing an ace on the first drawn then

$P(B_2 \mid A_1) = \frac{4}{51}$ and $P(B_1) = \frac{4}{52}$

again $P(B_2 \mid A_1) \neq P(B_1)$ since A and B are dependent events.

In throwing a pair of dice, let $P(A)$ is the probability of the event A that first die turns up odd and $P(B)$ is the probability of the event B that the second die turns up odd.

$$P(A) = \frac{1}{2}, \qquad P(B) = \frac{1}{2}$$

If C is the event that the total number of spots is odd, then C can occur only if the second dice turns up even, hence

$$P(C \mid A) = \frac{1}{2}$$

and similarly $P\,(C \mid B) = \frac{1}{2}$.

It follows $P\,(C \mid A) = P\,(C)$, $P\,(C \mid B) = P\,(C)$. Therefore the events A and C are independent, and so are the events B and C.

WORKED EXAMPLE 53

If $P\,(A) = 0.5$, $P\,(B) = 0.4$.

Find (i) $P\,(A \cap B)$ (ii) $P\,(A \cup B)$ (iii) $P\,(A \mid B)$ (iv) $P\,(A \cap \overline{B})$.

SOLUTION 53

(i) $P\,(A \cap B) = P\,(A) \times P\,(B) = 0.5 \times 0.4 = 0.2$

(ii) $P\,(A \cup B) = P\,(A) + P\,(B) - P\,(A \cap B)$
$= 0.5 + 0.4 - 0.2 = 0.9 - 0.2 = 0.7$

(iii) $P\,(A \mid B) = \dfrac{P\,(A \cap B)}{P\,(B)} = \dfrac{0.2}{0.4} = \dfrac{1}{2}$.

(iv) $P\,(A \cap \overline{B}) = P\,(A) \times P\,(\overline{B}) = P\,(A)\,[1 - P\,(B)]$
$= P\,(A) - P\,(A) \times P\,(B)$
$= 0.5 - 0.2 = 0.3$.

WORKED EXAMPLE 54

If $P\,(A \cap B) = P\,(A) \times P\,(B)$, show that A and B are independent events.

SOLUTION 54

$P\,(A \cap B) = P\,(A) \times P\,(B \mid A)$

A and B independent

$$P\,(B \mid A) = P\,(B \mid \overline{A}) = P\,(B) \Rightarrow P\,(A \cap B) = P\,(A) \times P\,(B).$$

Statement A and B are independent implies statement $P\,(A \cap B) = P\,(A) \times P\,(B)$.

$$P(A \cap B) = P(A) \times P(B) \Rightarrow P(B) = \frac{P(A \cap B)}{P(A)} = P(B \mid A)$$

$$P(B \mid \bar{A}) = \frac{P(B \cap \bar{A})}{P(\bar{A})} = \frac{P(B) - P(A \cap B)}{1 - P(A)}$$

$$= \frac{P(B) - P(A) \times P(B)}{1 - P(A)} = P(B)$$

$= P(B \mid A)$ so B is independent of A.

Statement A and B are independent implies statement $P(A \cap B) = P(A) \times P(B)$ and vice-versa.

WORKED EXAMPLE 55

From a pack of 52 playing cards we draw two successive cards

(i) without replacing the first card,
(ii) with replacing the first card.

Find the probability in each case for the cards to be 2 Queens.

SOLUTION 55

(i) A is the event of the first card to be a queen

B is the event of the second card to be a queen

Drawing $P(A \cap B) = P(A)\, P(B \mid A)$

$P(A) = \frac{4}{52}$, $P(B \mid A) = \frac{3}{51}$, $P(A \cap B) = \frac{4}{52} \cdot \frac{3}{51} = \frac{1}{221}$

(ii) $P(A) = \frac{4}{52}$, $P(B \mid A) = P(B) = \frac{4}{52}$, $P(A \cap B) = \frac{4}{52} \cdot \frac{4}{52} = \frac{1}{169}$

WORKED EXAMPLE 56

A box contain 4 red marbles and 3 white marbles. We draw a marble at random and we note its colour and do not replace it, then we draw another marble. What is the probability (a) the two marbles are white
(b) the two marbles are red.

SOLUTION 56

(i)

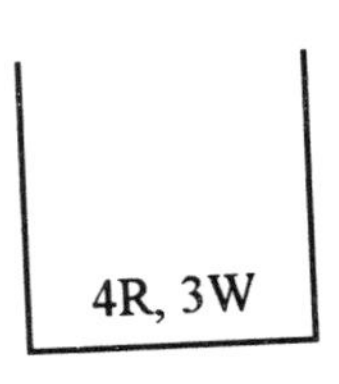

Fig. 10-I/23

$P\ (A)$ is the probability for the first marble to be white

$P\ (B)$ is the probability for the second marble to be white

$$P\ (A) = \frac{3}{7}$$

$P\ (B \mid A)$ = the probability of the second marble to be white given that the first marble is also white

$$= \frac{2}{6} = \frac{1}{3}$$

$$P\ (A \cap B) = \frac{3}{7} \cdot \frac{1}{3} = \frac{1}{7}$$

(ii) $P\ (A)$ is the probability for the first marble to be red

$P\ (B)$ is the probability for the second marble to be red

$P\ (B \mid A)$ = the probability of the second marble to be red given that A is also red.

$$P\ (A \cap B) = P\ (B \mid A)\ P\ (A) = \frac{3}{6} \cdot \frac{4}{7} = \frac{2}{7}$$

where $P\ (A) = \frac{4}{7}$

and $P\ (B \mid A) = \frac{3}{6} = \frac{1}{2}$.

Consider a pack of 52 cards of which 26 cards are red and 26 cards are black. If the first card is drawn at random and not replaced, then the possibility space for the drawing of a second card has been reduced by one element, the events are dependent.

If the second card is black (B) we have two different probabilities to consider.

$$P\ (B \text{ given that the first card removed is black}) = \frac{25}{51}$$

$$P\ (B \text{ given that the first card removed is red}) = \frac{26}{51}$$

$$P\left(B_2 \mid B_1\right) = \frac{25}{51} \quad \text{and} \quad P\left(B_2 \mid R_1\right) = \frac{26}{51}$$

where the subscript 2 denotes the second draw and the subscript 1 the first draw. The probability of the second draw depends on the first draw. The events are obviously dependent.

Consider the probability of drawing a red card first and a black card second, the

probability of drawing a red card first $P\left(R_1\right) = \frac{26}{52} = \frac{1}{2}$ and $P\left(B_2 \mid R_1\right) = \frac{26}{51}$

is the probability of drawing a black card second given that the first card removed is red.

In drawing a red card first and a black card second, there are 26×26 ways and there are 52×51 ways drawing these cards respectively.

$$P\left(R_1 \cap B_2\right) = \frac{26 \times 26}{52 \times 51} = \frac{1}{2} \times \frac{26}{51} = P\left(R_1\right) \times P\left(B_2 \mid R_1\right).$$

In general $P\ (A \cap B) = P\ (A) \times P\ (B \mid A)$. This is called the <u>compound probability</u> of both A and B occurring.

Therefore $\boxed{P\ (A \cap B) = P\ (A) \times P\ (B \mid A)}$ is a rule of probability

A and B are <u>dependent events</u>.

If A and B are independent events

$$P\ (A \cap B) = P\ (A) \times P\ (B) \text{ where } P\ (B \mid A) = P\ (B)$$

$A \cap B$ means (A AND B) ... (1)

$A \cup B$ means (A OR B) ... (2)

Equation (1) can be illustrated by a simple circuit of two switches A and B in series.

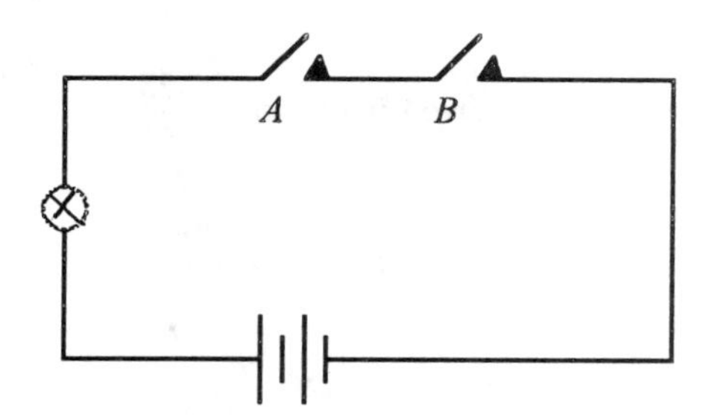

Fig. 10-I/24 $A \cap B$ (A AND B)

$$P(A \cap B) = P(A) \times P(B)$$

If switch A AND switch B are closed, there is continuity in the circuit and current flows and the light bulb will light.

Equation (2) can also be illustrated by a simple circuit of two switches A and B in parallel.

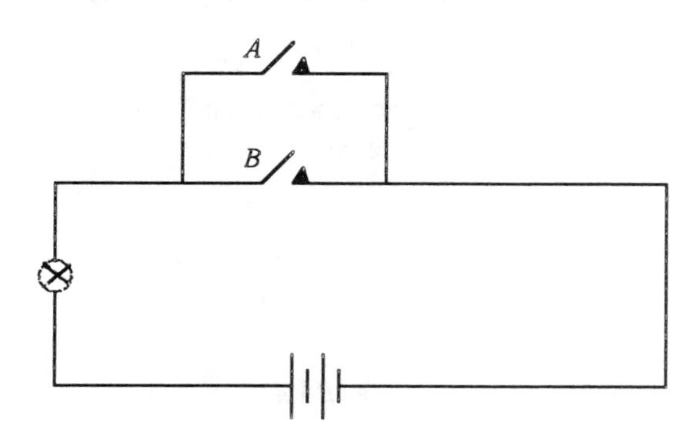

Fig. 10-I/25 $A \cup B$ (A OR B)

If switch A or switch B is closed, there will be continuity in the circuit and current flows and the light bulb will light.

$$P(A \cup B) = P(A) + P(B)$$

A and B are mutually exclusive

If A and B are not mutually exclusive events

P (A or B or both) $= P(A) + P(B) - P(A \text{ and } B)$

$$\boxed{P(A \cup B) = P(A) + P(B) - P(A \cap B)}$$

WORKED EXAMPLE 57

Two boxes are marked X and Y. Box X has 4 red discs and 3 white discs, box Y has 5 red discs and 4 white discs. A coin is now tossed, if the outcome is a head (H) we draw one disc from Box X, if the outcome is a tail (T) we draw one disc from Box Y.

Find the probability that the disc is white.

SOLUTION 57

Fig. 10-I/26

W is the event that the disc is white

Ax is the event that the disc comes from box X

Ay is the event that the disc comes from box Y.

The $P(Ax) = P(Ay) = \frac{1}{2}$

$P(W|Ax) = \frac{3}{7}$ $\qquad$ $P(W|Ay) = \frac{4}{9}$

$$P(W) = P(W \cap Ax) + P(W \cap Ay) = P(Ax)\,P(W|Ax) + P(Ay)\,P(W|Ay)$$

$$= \frac{1}{2} \cdot \frac{3}{7} + \frac{1}{2} \cdot \frac{4}{9} = \frac{3}{14} + \frac{4}{18}$$

$$= \frac{54 + 56}{14 \times 18} = \frac{110}{252} = \frac{55}{126}.$$

WORKED EXAMPLE 58

Consider the circuit show in Fig. 10-I/27

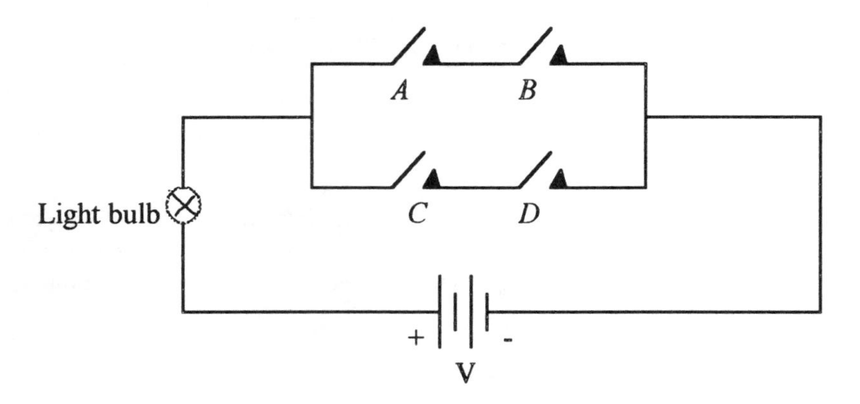

Fig. 10-I/27

The light bulb will light if A and B or C and D switch is closed when there will be continuity in the circuit.

If the probability of each switch to make (or to close) is P, given that all the switches operate independently. Find in term of P the probability for continuity in the circuit (the light bulb will light). If $P = \frac{1}{2}$, find the numerical probability.

SOLUTION 58

Let A, B, C and D be the events that each of the switches is closed and E be the event that there is a continuity (the light bulb will light).

Then $E = (A \cap B) \cup (C \cap D)$ and therefore

$$P(E) = P(A \cap B) + P(C \cap D) - P(A \cap B \cap C \cap D)$$

$$P(E) = P(A)\,P(B) + P(C)\,P(D) - P(A)\,P(B)\,P(C)\,P(D)$$

since the events A, B, C and D are independent.

$P(A) = P(B) = P(C) = P(D) = p$

$$P(E) = p^2 + p^2 - p^4 = 2p^2 - p^4$$

If $P = \frac{1}{2}$

$$P(E) = 2\left(\frac{1}{2}\right)^2 - \left(\frac{1}{2}\right)^4 = 2\left(\frac{1}{4}\right) - \frac{1}{16} = \frac{1}{2} - \frac{1}{16} = \boxed{\frac{7}{16}}$$

WORKED EXAMPLE 59

Consider the circuit shown in Fig. 10-I/28

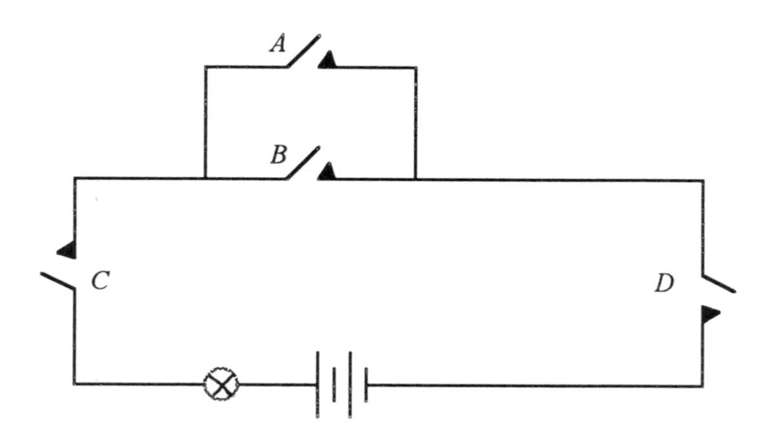

Fig. 10-I/28

The light bulb will light if (A or B) and C and D are closed when there will be continuity in the circuit.

If the probability of each switch to make (or to close) is $\frac{1}{4}$. Given that all the switches operate independently, find the probability for continuity in the circuit (the light bulb will light).

SOLUTION 59

Let A, B, C and D be the events that each of the switches is closed and E be the event that there is a continuity (the light bulb will light).

The $E = (\boldsymbol{A} \cup \boldsymbol{B}) \cap \boldsymbol{C} \cap \boldsymbol{D}$

therefore $\boldsymbol{P}(\boldsymbol{E}) = \boldsymbol{P}(\boldsymbol{A} \cup \boldsymbol{B}) \cap \boldsymbol{P}(\boldsymbol{C}) \cap \boldsymbol{C}(\boldsymbol{D})$

$$= [P(A) + P(B) - P(A)P(B)]\, P(C)\, P(D)$$

$$= \left[\frac{1}{4} + \frac{1}{4} - \frac{1}{4}\frac{1}{4}\right]\frac{1}{4}\cdot\frac{1}{4} = \left(\frac{1}{2} - \frac{1}{16}\right)\frac{1}{4}\cdot\frac{1}{4}$$

$$= \frac{7}{16} \times \frac{1}{16} = \frac{7}{256}.$$

WORKED EXAMPLE 60

We throw a dice 5 times. What is the probability

(i) of obtaining five 6.
(ii) of not obtaining a 6.
(iii) of obtaining at least one 6.

SOLUTION 60

(i) Let A_1, A_2, A_3, A_4 and A_5 be the events of the first, second, third, fourth, and fifth throw of obtaining a six, which are independent.

$$P(A) = P(A_1)\, P(A_2)\, P(A_3)\, P(A_4)\, P(A_5)$$

$$= P(A_1 \cap A_2 \cap A_3 \cap A_4 \cap A_5)$$

$$= \frac{1}{6}\cdot\frac{1}{6}\cdot\frac{1}{6}\cdot\frac{1}{6}\cdot\frac{1}{6} = \frac{1}{6^5} = \frac{1}{7776}$$

(ii) If B is the event of not obtaining six

$$P(B) = P\left(A_1' \cap A_2' \cap A_3' \cap A_4' \cap A_5'\right) = \frac{5}{6}\cdot\frac{5}{6}\cdot\frac{5}{6}\cdot\frac{5}{6}\cdot\frac{5}{6}$$

(note A' is the complement of A)

$$= \left(\frac{5}{6}\right)^5 = \frac{3125}{7776}$$

(iii) The probability of obtaining at least one six is 1 − P (no six)

$$P(B') = 1 - \frac{3125}{7776} \text{ for one six in five throws}$$

$$= \frac{4651}{7776}.$$

WORKED EXAMPLE 61

A dice is thrown twice

(i) Find the probability of throwing two fives, $P(A)$ and that the sum of the scores on the two dice is $7P(B)$.

(ii) Find the probability $P(C)$ that at least one of the throws is a five, and the conditional probabilities $P(A \mid C)$ and $P(B \mid C)$.

SOLUTION 61

(i) The probability of throwing firstly a five is $\frac{1}{6}$ and the probability of throwing secondly a five is also $\frac{1}{6}$, therefore the probability of throwing two fives is $\frac{1}{6} \times \frac{1}{6}$. Let $P(A)$ be the event of throwing two consecutive fives, and $P(B)$ be the event of throwing a sum of 7.

The possible ways of getting a score of 7 from two dice is (4, 3; 3, 4; 2, 5; 5, 2; 6, 1; 1, 6)

$$P(B) = \frac{6}{36} = \frac{1}{6}$$

(ii) Let $P(C)$ be the event that at least one of the throws is a five.

The number of ways of obtaining this is.

$C = \{(1, 5), (2, 5), (3, 5), (4, 5), (5, 5), (6, 5), (5, 1), (5, 2), (5, 3), (5, 4), (5, 6)\}$

$P(C) = \frac{11}{36}$ $\quad A \cap C = \{5, 5\}$, $\quad B \cap C = \{(5, 2), (2, 5)\}$

$$P(A \mid C) = \frac{P(A \cap C)}{P(C)} = \frac{1/36}{11/36} = \frac{1}{11}$$

$$P(B \mid C) = \frac{P(B \cap C)}{P(C)} = \frac{2/36}{11/36} = \frac{2}{11}.$$

WORKED EXAMPLE 62

We have two coloured dice, a green and a red. We toss a coin, if it lands 'head' we throw the green dice, if it lands 'tails' we throw the red dice.

What is the probability of obtaining a six?

SOLUTION 62

Let G be the event of throwing the green dice and R be the event of throwing the red dice, and A is the event of a six.

$$(A \cap G) \cap (A \cap R) = \emptyset$$

$$(A \cap G) \cup (A \cap R) = A$$

then we have

$$P(A) = P(A \cap G) + P(A \cap R) = P(A \mid G)\,P(G) + P(A \mid R)\,P(R)$$

$$= \frac{1}{6} \cdot \frac{1}{2} + \frac{1}{6} \cdot \frac{1}{2} = \frac{1}{12} + \frac{1}{12} = \frac{1}{6}.$$

Consider two unbiased dice, one yellow (number 1 to 6 with black dots) and one green (numbered 1 to 6 with white dots).

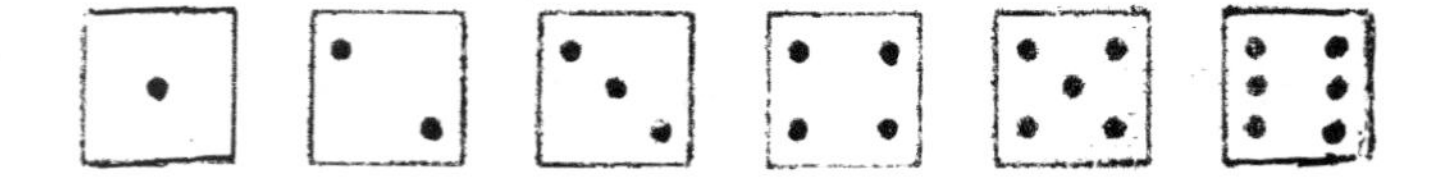

Fig. 10-I/29

The two dice are thrown together, what is the probability that they show a four on the green dice and an odd number on the yellow dice (one or three or five).

	1	2	3	4	5	6
1	X	X	X	X	X	X
2	X	X	X	X	X	X
3	X	X	X	X	X	X
4	(X)	X	(X)	X	(X)	X
5	X	X	X	X	X	X
6	X	X	X	X	X	X

YELLOW DICE

GREEN DICE

There are 36 equiprobable outcomes which are marked by an X as shown in the table. The probability that one die shows a four on the green dice and an odd number on the yellow dice $= \frac{1}{36} + \frac{1}{36} + \frac{1}{36} = \frac{3}{36} = \frac{1}{12}$.

This is rather tedious method, a much briefer method is to work out the probability of obtaining a four on the green dice is $\left(\frac{1}{6}\right)$ and the probability of obtaining an odd number on the yellow dice is $\left(\frac{1}{2}\right)$, therefore

P (4 on the green dice and an odd number on the yellow dice) $= \frac{1}{6} \times \frac{1}{2} = \frac{1}{12}$.

SIMPLE THREE DIAGRAMS

The previous example can be easily illustrated by a simple tree diagram.

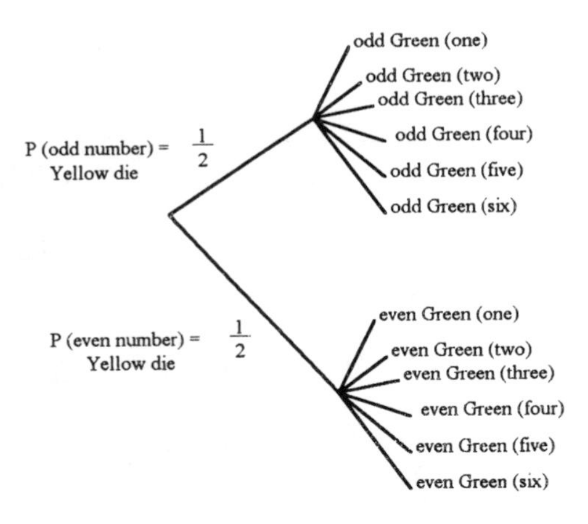

Fig. 10-I/30

There are twelve equiprobable outcomes, the probability required is $\frac{1}{12}$.

POSSIBILITY SPACE

In the set notation, the universal set, $\mathscr{E}$, for the example of the two dice, consists of 36 elements, each of which is equiprobable. This is call the possibility space.

$n(A) = 3$, $n(\mathscr{E}) = 36$, and $P(A) = \dfrac{n(A)}{n(\mathscr{E})}$

If A is the set of all outcomes favourable for one event, and B the set of all outcomes favourable for another, $A \cap B$ is the set of all outcomes favourable for both

$$P(A \cup B) = P(A) + P(B) - P(A \cap B)$$

TREE DIAGRAM OR PROBABILITY TREE

Another concept of keeping the ideas clear or straight is the tree diagram or the probability tree.

What are all the equally likely outcomes in tossing four an unbiased coin.

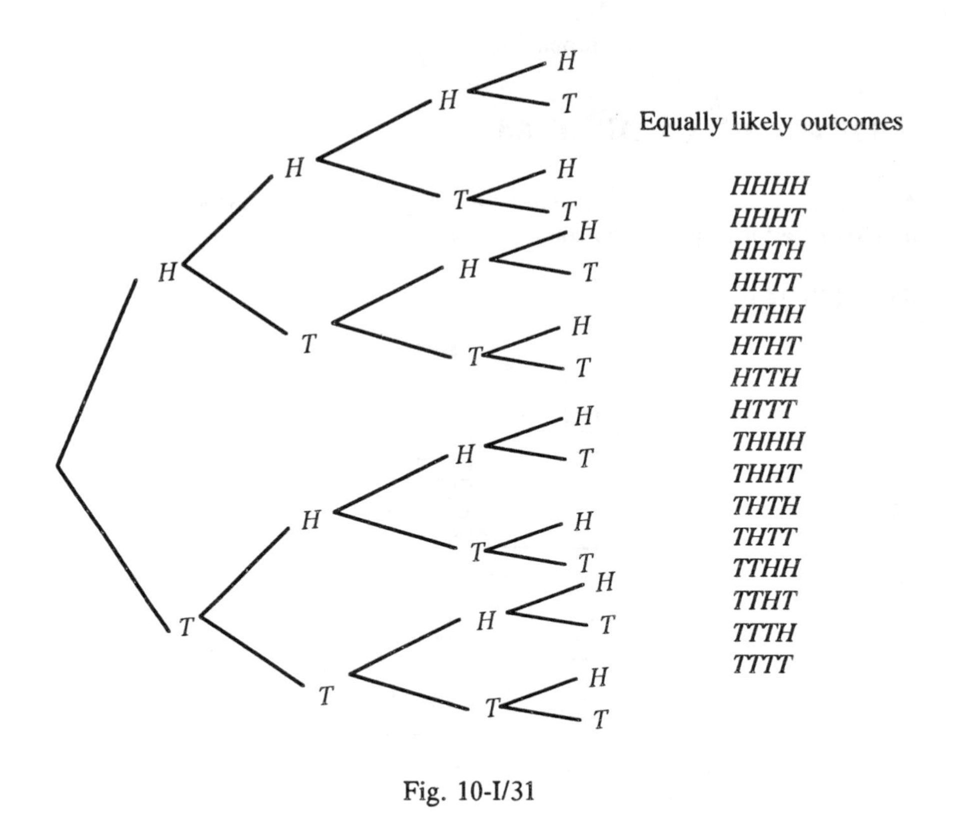

Fig. 10-I/31

The probability of throwing four heads in a row is $\frac{1}{16}$.

WORKED EXAMPLE 63

Use the probability tree above to answer the following questions:
What is the probability of getting: (i) Three tails in a row and a head?
(ii) One tail and three head in that order?
(iii) Three heads in any order?

SOLUTION 63

(i) There is only one way of getting *TTTH* and one way of getting *HTTT*, so the probability is $\frac{2}{16}$

(ii) There is only one way of getting *THHH*, so the probability is $\frac{1}{16}$

(iii) There are four ways of getting three heads in any order, *HHHT, HHTH, HTHH, THHH,* so the probability is $\frac{4}{16} = \frac{1}{4}$.

WORKED EXAMPLE 64

A game consists of tossing a coin and throwing two dice. Draw a probability tree, illustrating all the possible outcomes.

SOLUTION 64

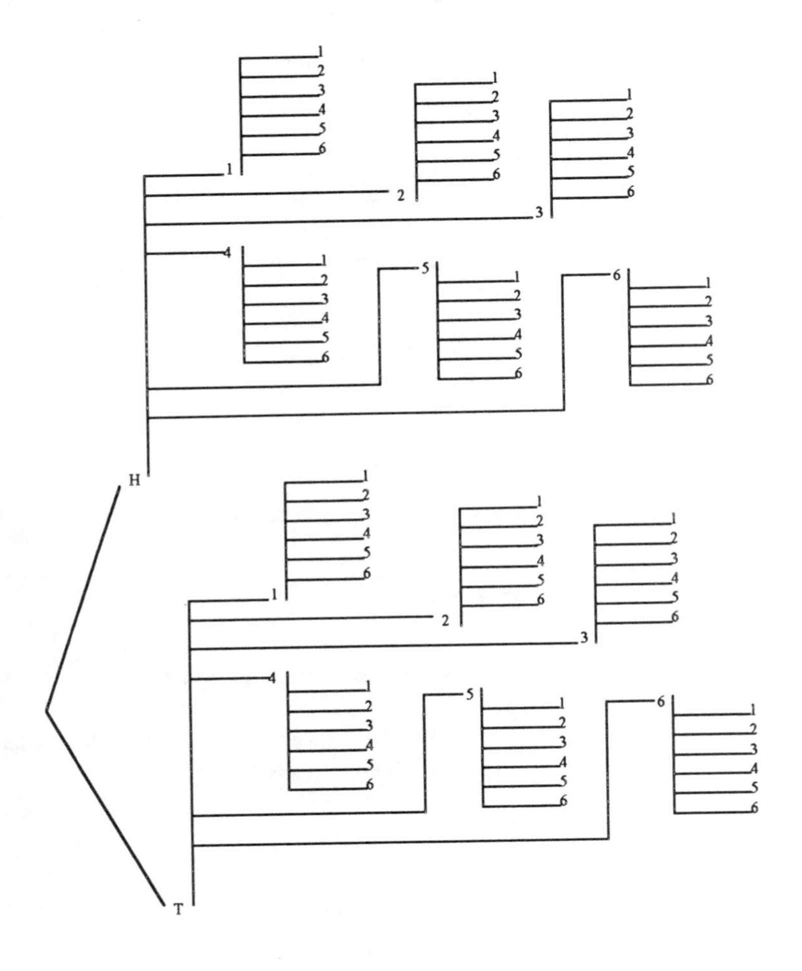

Fig. 10-I/32

There are 72 different ways

WORKED EXAMPLE 65

Three balls are drawn from a bag containing 4 reds, 5 yellow, 6 blue ones, the balls are identical except for the colour. Find the probabilities of drawing (i) 3 reds (ii) 3 yellows (iii) 3 blues (iv) 1 of each colour if (a) each ball is replaced after drawing and (b) the balls are not replaced.

SOLUTION 65

(a) (i) $P\,(1R) = \dfrac{4}{15}$, $\quad P\,(3R) = \left(\dfrac{4}{15}\right)^3 = \dfrac{64}{3375}$

(ii) $P\,(1Y) = \dfrac{5}{15} = \dfrac{1}{3}$, $\quad P\,(3Y) = \left(\dfrac{1}{3}\right)^3 = \dfrac{1}{27}$

(iii) $P\,(1B) = \dfrac{6}{15} = \dfrac{2}{5}$, $\quad P\,(3B) = \left(\dfrac{2}{5}\right)^3 = \dfrac{8}{125}$

(iv) $P\,(1R,\ 1Y,\ 1B) = \dfrac{3!}{1!\ 1!\ 1!} \times \dfrac{4}{15} \times \dfrac{1}{3} \times \dfrac{2}{5} = \dfrac{48}{225} = \dfrac{16}{75}$

Since there are $\dfrac{3!}{1!\ 1!\ 1!} = 6$ ways of selecting balls of different

colours. (*RYB*; *RBY*; *YRB*; *YBR*; *BYR*; *BRY*)

(b) (i) $P\,(1R) = \dfrac{4}{15}$, there are now 14 balls left with $3R$,

P (Second red) $= \dfrac{3}{14}$, there are now 13 balls left with $2R$,

P (third red) $= \dfrac{2}{13}$

$P\,(3R) = \dfrac{4}{15} \times \dfrac{3}{14} \times \dfrac{2}{13} = \dfrac{4}{13 \times 35} = \dfrac{4}{455}$

(ii) P (First yellow) $= \frac{5}{15} = \frac{1}{3}$, there are now 14 balls left with 4 yellow,

P (Second yellow) $= \frac{4}{14} = \frac{2}{7}$, P (Third Y) $= \frac{3}{13}$

$$P(3Y) = \frac{1}{3} \times \frac{2}{7} \times \frac{2}{13} = \frac{2}{91}$$

(iii) P (first B) $= \frac{6}{15} = \frac{2}{5}$

P (second B) $= \frac{5}{14}$

P (third B) $= \frac{4}{13}$

$$P(3B) = \frac{2}{5} \times \frac{5}{14} \times \frac{4}{13} = \frac{4}{91}$$

(iv) P (1 red) $= \frac{4}{15}$

P (1 yellow) $= \frac{5}{14}$

P (1 blue) $= \frac{6}{13}$

$$P(1R, 1Y, 1B) = \frac{3!}{1!\ 1!\ 1!} \times \frac{4}{15} \times \frac{5}{14} \times \frac{6}{13} = \frac{24}{91}$$

since there are $\frac{3!}{1!\ 1!\ 1!} = 6$ ways of selecting one red, one yellow and one blue ball.

WORKED EXAMPLE 66

An unbiased coin is tossed twice. Find the probability of landing the following:

(i) two heads (H),
(ii) two tails (T),
(iii) the second toss is a tail, given that the first toss is head $P(T_2 \mid H_1)$

(iv) the second toss is a head, given that the first toss is tail $P(H_2 \mid T_1)$.

SOLUTION 66

(i) $P(H_1) = \frac{1}{2}, \quad P(H_2) = \frac{1}{2}$

$$P(H_1 \cap H_2) = P(H_1) \times P(H_2) = \frac{1}{2} \times \frac{1}{2} = \frac{1}{4}$$

(ii) $P(T_1) = \frac{1}{2}, \quad P(T_2) = \frac{1}{2}$

$$P(T_1 \cap T_2) = P(T_1) \times P(T_2) = \frac{1}{2} \times \frac{1}{2} = \frac{1}{4}$$

(iii) $P(T_2 \mid H_1) = \frac{1}{2}$, the second toss has no bearing on the first

(iv) $P(H_2 \mid T_1) = \frac{1}{2}$, the second toss has no bearing on the first.

Observe that each second toss is independent from the first toss. The events are independent.

WORKED EXAMPLE 67

A perfect roulette wheel has 36 numbers, of which 18 are red and 18 are black and are placed round the wheel alternatively. A ball is spun twice. Find the probability of the following outcomes:-

(i) two reds (R)
(ii) two blacks (B)
(iii) the second spin is a red, given that the first spin is also red, $P(R_2 \mid R_1)$
(iv) the second spin is a black, given that the first spin is red, $P(B_2 \mid R_1)$
(v) the second spin is a red, given that the first spin is black, $P(R_2 \mid B_1)$
(vi) the second spin is a black, given that the first spin is black, $P(B_2 \mid B_1)$
(vii) the second spin is a red, given that the first spin is black, $P(R_2 \mid B_1)$.

SOLUTION 67

(i) $P(R_1) = \frac{1}{2}, \; P(R_2) = \frac{1}{2}; \; P(R_1 \cap R_2) = P(R_1) \times P(R_2) = \frac{1}{2} \times \frac{1}{2} = \frac{1}{4}$

(ii) $P(B_1) = \frac{1}{2}, \qquad P(B_2) = \frac{1}{2};$

$$P(B_1 \cap B_2) = P(B_1) \times P(B_2) = \frac{1}{2} \times \frac{1}{2} = \frac{1}{4}$$

(iii) $P(R_2 \mid R_1) = \frac{1}{2}$, the second spin has no influence on the first spin

(iv) $P(B_2 \mid R_1) = \frac{1}{2}$, the second spin has no influence on the first spin

(v) $P(R_2 \mid B_1) = \frac{1}{2}$, the second spin has no influence on the first spin

(vi) $P(B_2 \mid B_1) = \frac{1}{2}$, the second spin has no influence on the first spin

(vii) $P(R_2 \mid B_1) = \frac{1}{2}$, the second spin has no influence on the first spin.

WORKED EXAMPLE 68

An unmarked pack of 52 cards has 26 black cards and 26 red cards. The pack is properly shuffled and two cards are drawn at random. Find the probability of turning up the following:

(i) two red cards (R),
(ii) two black cards (B),
(iii) the second card is red, given that the first card is also red $P(R_2 \mid R_1)$
(iv) the second card is red, given that the first card is black $P(R_2 \mid B_1)$
(v) the second card is black, given that the first card is black $P(B_2 \mid B_1)$,
(vi) the second card is black, given that the first card is red $P(B_2 \mid R_1)$.

when;
(a) The first card in each case is not replaced.
(b) the first card in each case is replaced.

SOLUTION 68

(a) (i) $P(R_1) = \frac{26}{52} = \frac{1}{2}, \; P(R_2 \mid R_1) = \frac{25}{51},$

$$P(R_1 \cap R_2) = P(R_1) \times P(R_2 \mid R_1) = \frac{1}{2} \times \frac{25}{51} = \frac{25}{102}$$

(ii) $P(B_1) = \frac{26}{52} = \frac{1}{2}, \quad P(B_2 \mid B_1) = \frac{25}{51},$

$$P(B_1 \cap B_2) = P(B_1) \times P(B_2 \mid B_1) = \frac{1}{2} \times \frac{25}{51} = \frac{25}{102}$$

(iii) $P(R_2 \mid R_1) = \frac{25}{51}$

(iv) $P(R_2 \mid B_1) = \frac{26}{51}$

(v) $P(B_2 \mid B_1) = \frac{25}{51}$

(vi) $P(B_2 \mid R_1) = \frac{26}{51}$

(b) (i) $P(R_1) = \frac{1}{2}, P(R_2) = \frac{1}{2}, \quad P(R_1 \cap R_2) = P(R_1) \times P(R_2) = \frac{1}{4}$

(ii) $P(B_1) = \frac{1}{2}, P(B_2) = \frac{1}{2}, \quad P(B_1 \cap B_2) = P(B_1) \times P(B_2) = \frac{1}{4}$

(iii) $P(R_2) = \frac{1}{2}$

(iv) $P(R_2) = \frac{1}{2}$

(v) $P(B_2) = \frac{1}{2}$

(vi) $P(B_2) = \frac{1}{2}$.

Comment.

For (a) the events are dependent.
For (b) the events are independent.

WORKED EXAMPLE 69

An unbiased dodecahedron (twelve faces) die, marked 1–12 is rolled twice. Find the probabilities of the following:

(i) rolling two tens,
(ii) rolling two twelves,
(iii) the second throw is a seven, given that the first throw is a seven,
(iv) obtaining a score of 24 from the two throws,
(v) not obtaining a twelve, in two throws,
(vi) obtaining a score of 12 from the two throws,
(vii) obtaining at least one, '1' from two throws,
(viii) obtaining only one seven from two throws.

SOLUTION 69

(i) $P\left(10_1\right) = \frac{1}{12}, P\left(10_2\right) = \frac{1}{12}, \qquad P\left(10_1 \cap 10_2\right) = \frac{1}{12} \times \frac{1}{12} = \frac{1}{144},$

(ii) $P\left(12_1\right) = \frac{1}{12}, P\left(12_2\right) = \frac{1}{12}, \qquad P\left(12_1 \cap 10_2\right) = \frac{1}{12} \times \frac{1}{12} = \frac{1}{144},$

(iii) $P\left(7_2 \mid 7_1\right) = \frac{1}{12} = P\left(7_2\right) = \frac{1}{12},$

(iv) $P\left(12_1 \cap 12_2\right) = \frac{1}{12} \times \frac{1}{12} = \frac{1}{144},$

(v) The probability of not getting a twelve on one throw is $\frac{11}{12}$.

$$P\left(\overline{12_1} \cap \overline{12_2}\right) = P\left(\overline{12_1}\right) \times P\left(\overline{12_2}\right) = \frac{11}{12} \times \frac{11}{12} = \frac{121}{144}$$

$$\boxed{P\left(\bar{A}\right) = \frac{121}{144}}$$

where A is the event of throwing of at least one twelve, $\bar{A}$ is not throwing twelve on two throws.

(vi) To obtain a score of 12 from the two throws,

$\left(1_1 \cap 11_2\right), \left(2_1 \cap 10_2\right), \left(3_1 \cap 9_2\right), \left(4_1 \cap 8_2\right), \left(5_1 \cap 7_2\right), \left(6_1 \cap 6_2\right),$

$\left(11_1 \cap 1_2\right), \left(10_1 \cap 2_2\right), \left(9_1 \cap 3_2\right), \left(8_1 \cap 4_2\right), \left(7_1 \cap 5_2\right).$

$$P\text{ (score of 12)} = P(1_1 \cap 11_2) + P(2_1 \cap 10_2) + P(3_1 \cap 9_2) + P(4_1 \cap 8_2) + P(5_1 \cap 7_2) + P(6_1 \cap 6_2) + P(11_1 \cap 1_2) + P(10_1 \cap 2_2) + P(9_1 \cap 3_2) + P(8_1 \cap 4_2) + P(7_1 \cap 5_2).$$

$$= 11 \times \left(\frac{1}{12} \times \frac{1}{12}\right) = \boxed{\frac{11}{144}} \text{ where}$$

$$P(1_1 \cap 11_2) = P(1_1) \times P(11_2) = \frac{1}{12} \times \frac{1}{12} = \frac{1}{144}$$

The scores from each dice are mutually exclusive events.

(vii) The probability of not obtaining a one is $\frac{11}{12}$,

$$P(\overline{1}_1 \cap \overline{1}_2) = P(\overline{1}_1) \times P(\overline{1}_2) = \frac{11}{12} \times \frac{11}{12} = \frac{121}{144}$$

$$P(A) = 1 - P(\overline{A}) = 1 - \frac{121}{144} = \frac{144 - 121}{144} = \frac{23}{144}$$

the probability of obtaining at least one, '1'.

(viii) To obtain only one seven, either the first throw is a seven and the second throw is not a seven or the first throw is not a seven and the second throw is a seven.

$$P(7_1 \cap \overline{7}_2) + P(\overline{7}_1 \cap 7_2) = P(7_1) \times P(\overline{7}_2) + P(\overline{7}_1) \times P(7_2)$$

$$= \frac{1}{12} \times \frac{11}{12} + \frac{11}{12} \times \frac{1}{12} = \frac{11}{144} + \frac{11}{144} = \boxed{\frac{11}{72}}$$

WORKED EXAMPLE 70

An unbiased coin is to be tossed n times. Calculate the probability of obtaining at least one tail.

SOLUTION 70

$$1 - \left(\frac{1}{2}\right)^n.$$

WORKED EXAMPLE 71

A fair coin is to be tossed four times. Calculate, the probability of obtaining at least one tail.

SOLUTION 71

$$1 - \left(\frac{1}{2}\right)^4 = 1 - \frac{1}{16} = \frac{15}{16}.$$

WORKED EXAMPLE 72

An unbiased coin is to be tossed 9 times. Calculate, correct to three places of decimals, the probability of obtaining at least one head.

SOLUTION 72

$$1 - \left(\frac{1}{2}\right)^9 = 0.998.$$

WORKED EXAMPLE 73

In a game of Roulette, a ball is span 11 times, where there are 18 red numbers and 18 black numbers. Calculate, as a rational number, the probability of obtaining at least one red number.

SOLUTION 73

$$1 - \left(\frac{1}{2}\right)^{11} = \frac{2047}{2048}.$$

WORKED EXAMPLE 74

(a) The following table shows the distribution by size and by colour of a batch of 550 dresses of similar design

Dresses	Size 12	14	16	18
Green	23	45	39	13
Blue	49	92	83	56
Pink	68	23	28	21

A dress is selected at random from this batch. If A is the event that a size 16 dress is selected, B is the event that the chosen dress is blue and C is the event that the chosen dress is green.

(i) Find $P(A)$, $P(B)$, $P(C)$ and $P(\overline{C})$

(ii) Find $P(B \cup C)$

(iii) Find $P(A \cup B)$

(iv) Find $P(A \cap C)$

(v) Find $P(C \mid A)$.

(b) If three dresses were selected at random from this batch without replacement, find the probability that only two pink were chosen.

SOLUTION 74

Redraw the table above and sum the total green dresses, the blue and pink dresses, total the size 12, 14, 16 and 18 as shown

Dresses	Sizes				Total
	12	14	16 (A)	18	
Green (C)	23	45	39	13	120
Blue (B)	49	92	83	56	280
Pink	68	23	28	21	140
	140	160	150	90	540

Let A = size 16, B = Blue, C = Green.

(i) $P(A) = \frac{150}{540} = \frac{5}{18}$ $\qquad P(B) = \frac{280}{540} = \frac{14}{27}$

$P(C) = \frac{120}{540} = \frac{2}{9}$ $\qquad P(\overline{C}) = 1 - \frac{2}{9} = \frac{9-2}{9} = \frac{7}{9}$

(ii) $P(B \cup C) = P(B) + P(C) = \frac{280}{540} + \frac{120}{540} = \frac{20}{27}$

(iii) $P(A \cup B) = P(A) + P(B) - P(A \cap B) = \frac{150}{540} + \frac{280}{540} - \frac{83}{540} = \frac{347}{540}$

(iv) $P(A \cap C) = \frac{39}{550}$

(v) $P(C \mid A) = \frac{P(A \cap C)}{P(A)} = \frac{39/540}{150/540} = \frac{39}{150} = \frac{13}{50}$

(b) $P(2P \text{ and } 1\overline{P}) = P(ppp') = \frac{3!}{2!\ 1!} \times \frac{140}{540} \times \frac{139}{539} \times \frac{400}{538} = 0.149128$

WORKED EXAMPLE 75

A shop stocks 1 ampere, 5 ampere and 13 ampere fuses from two suppliers A and B. He has 100 fuses in stock of which 70% are from supplier A and 30% from supplier B. Altogether there are 60 × 13 ampere fuses in stock, and of these 40 are from supplier A. There are 20 × 5 ampere fuses in stock of which 15 were from supplier A.

(a) One fuse is selected at random.

Let A be the event the fuse is from supplier A
F be the event that the fuse selected is a 5 ampere fuse
T be the event that the fuse selected is a 13 ampere fuse.

Find (i) $P(A)$, $P(F)$, $P(T')$, $P(A \cap F)$

(ii) $P(A \cup F)$ and $P(F \cup T)$

(iii) $P(A \mid F)$

(iv) $P(A \cup T')$.

(b) If three fuses are randomly selected find the probability that only one is a 13 ampere fuse.

SOLUTION 75

SUPPLIER	AMPERES			Total
	13(T)	5 (F)	1	
A	40	15	15	70
B	20	5	5	30
	60	20	20	100

(i) $P(A) = \frac{70}{100} = \frac{7}{10}, \quad P(F) = \frac{20}{100} = \frac{1}{5}, \quad P(T') = \frac{40}{100} = \frac{2}{5}$

$P(A \cap F) = \frac{15}{100} = \frac{3}{20}.$

(ii) $P(A \cup F) = P(A) + P(F) - P(A \cap F)$

$$= \frac{70}{100} + \frac{20}{100} - \frac{15}{100} = \frac{75}{100} = \frac{3}{4}$$

$P(F \cup T) = P(F) + P(T) = \frac{60}{100} + \frac{20}{100} = \frac{80}{100} = \frac{4}{5}.$

(iii) $P(A \mid F) = \frac{P(A \cap F)}{P(F)} = \frac{15/100}{20/100} = \frac{15}{20} = \frac{3}{4}.$

(iv) $P(A \cup T') = P(A) + P(T') - P(A \cap T')$

$$= \frac{70}{100} + \frac{40}{100} - \frac{30}{100} = \frac{80}{100} = \frac{4}{5}.$$

(b) $P(T T' T') = \frac{3!}{2!\ 1!} \times P(T) \times P(T') \times P(T')$

$$= 3 \times \frac{60}{100} \times \frac{40}{99} \times \frac{39}{98}$$

$$= 3 \times \frac{3}{5} \times \frac{20}{99} \times \frac{39}{49} = \frac{20}{55} \times \frac{39}{49}$$

$$= \frac{156}{539}.$$

EXERCISES 3, 4 AND 5

1. Find the probability of throwing 5 consecutive heads with five throws of a coin.

2. Find the probability of throwing 3 consecutive tails with three throws of a coin.

3. The probability of drawing a king from a pack of cards is $\frac{1}{13}$ and the probability of drawing a spade is $\frac{1}{4}$. Find the probability of drawing the king of spades.

4. A bag contains 3 red balls, 4 yellow balls and 5 blue balls, all are identical. Find the probability of drawing 2 red balls and 1 blue ball, without replacing the balls.

5. A bag contains 5 red discs, 4 yellow discs and 3 blue discs, all are identical. Find the probability of drawing 1 red, 1 yellow and 1 blue disc, without replacing the discs.

6. A bag contains 3 balls, 1 Red, 1 Yellow and 1 Blue, all are identical. Find the probability of drawing first Blue ball then Yellow without replacing the balls.

7. Seventeen coins are tossed. What is the probability (i) of landing 7 heads, (ii) of landing 10 tails.

8. 8 coins are tossed find the probability of 6 heads appearing.

9. $P(A) = 0.6$, $P(B) = 0.7$.

   Find (i) $P(A \cap B)$

   (ii) $P(A \cup B)$.

10. $P(A) = 0.1$, $P(B) = 0.2$.

    Find (i) $P(A \cap B)$ (ii) $P(A \cup B)$.

11. If $P(A \cap B) = 0.1$ and $P(A \cup B) = 0.8$.

    Find (i) $P(A)$ (ii) $P(B)$.

12. If $P(A \cup B) = 0.7$ and $P(A \cap B) = 0.07$

    Find (i) $P(A)$ (ii) $P(B)$.

13. A biased die is constructed so that each of the numbers 1, 2 and 6 are twice as likely to occur as each of the other three numbers 3, 4 and 5. Find

(a) the probability of throwing a one,
(b) the probability of throwing a six,
(c) the probability of throwing a three,

14. A biased die is constructed so that each of the numbers 3 and 4 are thrice as likely to occur as each of the numbers 1, 2, 5 and 6. Find

(a) the probability of throwing a 3
(b) the probability of throwing a 4 given that the throw is greater than 3.

15. In a poker game with 32 playing cards A, 7, 8, 9, 10, J, Q, K. Find the probability of drawing (i) one ace (ii) two aces (iii) three aces and (iv) four aces.

16. In a poker game with 32 playing cards A, 7, 8, 9, 10, J, Q, K. Find the probability of drawing 3Q and 2J.

17. A hand of five cards is to be drawn without replacement and at random from a pack of 52 playing cards. Find the probabilities that this hand will contain:

(a) 4 aces and a king
(b) 3 kings and two queens

Given your answer in each case to three significant figures.

6. THE MATHEMATICS OF GAMBLING

THE GAME OF CRAPS

THE PASS LINE BET

Two unbiased coloured dice, one green and one yellow, each of six faces marked 1 - 6, hexahedrons. Let us denote the 36 equiprobable displays as follows:-

GY	*GY*	*GY*	*GY*	*GY*	*GY*
1,1	2,1	3,1	4,1	5,1	6,1
1,2	2,2	3,2	4,2	5,2	6,2
1,3	2,3	3,3	4,3	(5,3)	6,3
1,4	2,4	3,4	4,4	5,4	6,4
1,5	2,5	(3,5)	4,5	5,5	6,5
1,6	2,6	3,6	4,6	5,6	6,6

TABULATION OF ALL EQUALLY LIKELY OUTCOMES. POSSIBILITY SPACE. 36 SAMPLE POINTS

NOTATION

I have circled **3,5** and **5,3**, the first digits relate to the green die and the second digits relate to the yellow die.

To win, the player must first roll eleven or seven. The probability of winning on the first roll of the two dice is $\frac{2}{36} = \frac{1}{18}$ since there are two probable ways of obtaining eleven, 5 + 6 or 6 + 5, that is 5,6 or 6,5. The probability of winning on the first roll of the two dice with seven is $\frac{6}{36} = \frac{1}{6}$. Therefore to win on the first roll, the probability is $\frac{1}{18} + \frac{1}{6} = \frac{1}{18} + \frac{3}{18} = \frac{4}{18} = \frac{2}{9}$.

The player, however, loses if the two dice when rolled sum up to 2(1,1), 3(1,2 or 2,1) or 12 (6,6), the probability of losing is $\frac{1}{36} + \frac{1}{36} + \frac{1}{36} + \frac{1}{36} = \frac{4}{36} = \frac{1}{9}$.

The player will continue to play if he did not score 7, 11, 2, 3 or 12.

When the player throws the dice again and obtains the score that he obtained on the first throw, and before throwing a score of 7, he will win; otherwise he will lose if the score of 7 occurs before he obtained his original score.

Suppose the player rolls the score of 5 which can be achieved by the following (1,4; 2,3; 3,2; 4,1), he must throw 5 before 7, if seven turns up before 5, he loses, if he throws 5 before seven he wins. To roll seven there are 6 ways (6,1; 5,2; 4,3; 3,4; 2,5; 1,6) and to roll 5 there are four ways. Therefore he loses on six rolls and wins on four rolls. The conditional probability of winning is $\frac{4}{10} = \frac{2}{5}$.

The overall chance of winning by first throwing a five and then making the point before seven is rolled is given by multiplying $\frac{1}{9} \times \frac{2}{5} = \frac{2}{45}$.

Initial Score	Probability of total	Probability of winning with this score	Total probability of winning with this score
2	1/36 (1,1)	-	-
3	2/36 (1,2 : 2,1)	-	-
4	3/36 (1,3 : 3,1 : 2,2)	3/9	$\frac{3}{36} \times \frac{3}{9} = \frac{55}{1980}$
5	4/36 (1,4 : 4,1 : 2,3 : 3,2)	4/10	$\frac{4}{36} \times \frac{4}{10} = \frac{88}{1980}$
6	5/36 (2,4 : 4,2 : 3,3 : 1,5 : 5,1)	5/11	$\frac{5}{36} \times \frac{5}{11} = \frac{125}{1980}$
7	6/36 (6,1:5,2:4,3:3,4:2,5:1,6)	-	$\frac{6}{36} = \frac{330}{1980}$
8	5/36 (6,2 : 5,3 : 4,4 : 3,5 : 2,6)	5/11	$\frac{5}{36} \times \frac{5}{11} = \frac{125}{1980}$
9	4/36 (6,3 : 5,4 : 4,5 : 3,6)	4/10	$\frac{4}{36} \times \frac{4}{10} = \frac{88}{1980}$
10	3/36 (6,4 : 5,5 : 4,6)	3/9	$\frac{3}{16} \times \frac{3}{9} = \frac{55}{1980}$
11	2/36 (6,5 : 5,6)	-	$\frac{2}{36} = \frac{11}{1980}$
12	1/36 (6,6)	-	-
		Total probability of winning	$\frac{976}{1980} = \frac{244}{495}$

The LCM of 36, 9, 10, 11 is $3 \times 3 \times 2 \times 2 \times 5 \times 11 = 180 \times 11 = 1980$
The prime factors of each number

$$36 = 3 \times 2 \times 2 \times 3$$
$$9 = 3 \times 3$$
$$10 = 2 \times 5$$
$$11 = 11.$$

The probability of winning in the game of craps is

$\frac{976}{1980} = \frac{488}{990} = \frac{244}{495} = 0.4929292$. The probability of losing in the game of

craps is $1 - \frac{976}{1980} = \frac{1004}{1980} = 0.5070707$

For the "pass line" bet the expectation is

$$1 \times \frac{976}{1980} - 1 \times \frac{1004}{1980} = -\frac{28}{1980} = -1.41\%$$

Tossing or rolling a pair of dice, what is the probability of landing two sixes?

$\frac{1}{36}$. What is the probability of not landing two sixes, $\frac{35}{36}$.

$$\left(\frac{35}{36}\right)^x = \frac{1}{2}$$

where x is the times of tossing a pair of dice in order to have an even chance.

$$\ln\left(\frac{35}{36}\right)^x = \ln\frac{1}{2}$$

$$x = \frac{\ln 1/2}{\ln 35/36} = 24.605098 \approx 24.6$$

24 rolls is too few and 25 rolls is too many to have exactly an even chance.

$$\ln\left(\frac{35}{36}\right) = \ln\left(1 - \frac{1}{36}\right) = -\left(\frac{1}{36} + \frac{1}{36 \times 2} + \frac{1}{36 \times 3} + \dots\right)$$

$$x = \frac{-\ln 2}{\ln 35/36} = \frac{-\ln 2}{-1/36} = 36 \ln 2 \approx 24.95 \approx 25.$$

PERCENTAGE PROBABILITY

What is the probability for landing a head when tossing a coin? The answer is $\frac{1}{2}$ or 0.5 or 50%.

ODDS ON AND ODDS AGAINST AN EVENT HAPPENING

Throwing a die on the table, the probability of getting a six on the upper most face is $\frac{1}{6}$. The odds against is 5 to 1.

In general if the <u>odds against</u> an event happening are x to y then the chance of the event happening is $\frac{y}{x+y}$. If <u>the odds on</u> an event happening are x to y then the chance of the event happening is $\frac{x}{x+y}$.

WORKED EXAMPLE 76

There are fifteen balls in a bag, five white balls and ten black balls. If one ball is drawn, what are the odds against this being white?

SOLUTION 76

The probability of drawing a white ball is $\frac{5}{15} = \frac{1}{3}$, the probability of failing to draw

a white ball is $\frac{10}{15} = \frac{2}{3}$.

Hence, the odds against this event happening are $\frac{2/3}{1/3} = \frac{2}{1}$ or 2 to 1.

The probability of drawing a black ball is $\frac{10}{15} = \frac{2}{3}$ and the probability of failing to draw a black ball is $\frac{5}{15} = \frac{1}{3}\left(1 - \frac{2}{3}\right)$.

The odds against drawing a black ball is 1 to 2

WORKED EXAMPLE 77

During a fête a lottery is conducted with 1000 tickets. A man buys 25 tickets, what is the probability of losing the lottery.

SOLUTION 77

$$\frac{1000 - 25}{1000} = \frac{975}{1000} = 0.975.$$

USING BOTH ADDITION AND MULTIPLICATION RULES

There are m red balls and n black balls. Find the probabilities of drawing

(i) one red ball and then one black ball
(ii) one black ball and then one red ball
(iii) one red ball and then one black ball OR one black ball and then one red ball
(iv) one black ball and then one black ball
(v) one red ball and then one red ball
(vi) either one or another of the four possible results [(i), (ii), (iv) or (v)] will take place

Solution

(i) $$\frac{m}{m+n} \times \frac{n}{m+n-1} = \frac{mn}{(m+n)(m+n-1)}$$

(ii) $$\frac{n}{m+n} \times \frac{m}{m+n-1} = \frac{nm}{(m+n)(m+n-1)}$$

(iii) $$\frac{mn}{(m+n)(m+n-1)} + \frac{mn}{(m+n)(m+n-1)} = \frac{2mn}{(m+n)(m+n-1)}$$

(iv) $$\frac{n}{m+n} \times \frac{n-1}{m+n-1} = \frac{n(n-1)}{(m+n)(m+n-1)}$$

(v) $$\frac{m}{m+n} \times \frac{m-1}{m+n-1} = \frac{m(m-1)}{(m+n)(m+n-1)}$$

(vi) (i) + (ii) + (iv) + (v) = $$\frac{mn}{(m+n)(m+n-1)} + \frac{mn}{(m+n)(m+n-1)} + \frac{n(n-1)}{(m+n)(m+n-1)} + \frac{m(m-1)}{(m+n)(m+n-1)}$$

$$= \frac{mn + mn + n^2 - n + m^2 - m}{(m + n)(m + n - 1)}$$

$$= \frac{n^2 + m^2 + 2mn - (n + m)}{(m + n)^2 - (m + n)} = 1.$$

Therefore one of the events B then R, or R then B or B then B or R then R is certain to happen.

WORKED EXAMPLE 78

In a bag there are 7 yellow (Y) balls and 6 blue (B) balls. Find the probabilities of drawing
(i) $1Y$ and then $1Y$
(ii) $1Y$ and then $1B$
(iii) $1B$ and then $1Y$
(iv) $1Y$ and then $1B$ or $1B$ and then $1Y$,
(v) $1B$ and then $1B$. Check your answer.

SOLUTION 78

(i) $\frac{7}{13} \times \frac{6}{12} = \frac{7}{26}$

(ii) $\frac{7}{13} \times \frac{6}{12} = \frac{7}{26}$

(iii) $\frac{6}{13} \times \frac{7}{12} = \frac{7}{26}$

(iv) $\frac{7}{13} \times \frac{6}{12} + \frac{6}{13} \times \frac{7}{12} = \frac{7}{26} + \frac{7}{26} = \frac{7}{13}$

(v) $\frac{6}{13} \times \frac{5}{12} = \frac{5}{26}$

(i) + (ii) + (iii) + (v) = $\frac{42}{156} + \frac{42}{156} + \frac{42}{156} + \frac{30}{156} = \frac{156}{156} = 1.$

WORKED EXAMPLE 79

A hand of four cards is to be drawn without replacement and at random from a pack of fifty two playing cards. Find the probabilities that this hand will contain
(i) two aces and two Kings
(ii) two aces and two Queens
(iii) two aces and two Kings OR two aces and two Queens.

SOLUTION 79

(i) $$\frac{4}{52}\times\frac{3}{51}\times\frac{4}{50}\times\frac{3}{49} = \frac{144}{49\times 50\times 51\times 52} = \frac{144}{6497400} = \frac{12}{541450}$$

(ii) $$\frac{4}{52}\times\frac{3}{51}\times\frac{4}{50}\times\frac{3}{49} = \frac{144}{49\times 50\times 51\times 52} = \frac{12}{541450}$$

(iii) $$\frac{288}{52\times 51\times 50\times 49} = \frac{12}{270725} = 0.44325\times 10^{-4}$$

MATHEMATICAL EXPECTATION

The expectation of a distribution X is simply the arithmetic mean of that distribution.

WORKED EXAMPLE 80

A person playing a game rolls 3 dice. If he throws 3 sixes he wins £60, if he throws 2 sixes he wins £30, if throws 1 six he wins £10, otherwise he loses £5. Find his expected winning for (i) each throw
(ii) ten throws.

SOLUTION 80

We need to evaluate the probabilities of winning and losing

$$P(6, 6, 6) = \frac{1}{6}\times\frac{1}{6}\times\frac{1}{6} = \frac{1}{216}$$

$$P(6, 6, \overline{6}) = \frac{3!}{2!\ 1!}\times\frac{1}{6}\times\frac{1}{6}\times\frac{5}{6} = \frac{5}{72}$$

$$P(6, \overline{6}, \overline{6}) = \frac{3!}{1!\ 2!}\times\frac{1}{6}\times\frac{5}{6}\times\frac{5}{6} = \frac{25}{72}$$

$$P(\overline{6}, \overline{6}, \overline{6}) = \frac{5}{6}\times\frac{5}{6}\times\frac{5}{6} = \frac{125}{216}.$$

(i) The expected winning per throw $= \sum_x xP_x$

$$= £60P(6,6,6) + £30P(6,6,\bar{6}) + £10P(6,\bar{6},\bar{6}) - £5P(\bar{6},\bar{6},\bar{6})$$

$$= £\left(60 \times \frac{1}{216} + 30 \times \frac{5}{72} + 10 \times \frac{25}{72} - 5 \times \frac{125}{216}\right)$$

$$= £\left(\frac{5}{18} + \frac{25}{12} + \frac{125}{36} - \frac{625}{216}\right)$$

$$= £2.94$$

(ii) The expected winning for 10 throws $= 10 \sum_x P_x$

$= 10\ £2.94$

$= £29.40$

An unbiased coin (a coin we assume will produce 50% heads and 50% tails)

$$(0.5)\ (1) + (0.5)\ (-1) = 0$$

If the probability of winning – the probability of losing is zero

$$P_w - P_L = 0$$

It is a fair game. The player has no advantage or disadvantage.

Suppose the pay off was changed to 3/2 (a gain of £1.50 in addition to our £ bet)

$$(0.5)\ (1.50) + (0.5)\ (-1) = 0.25.$$

Playing this game 100 times would give us a positive expectation of £25.

Consider the dozens bet in roulette our expectation for a £1 is:

$$\left(\frac{12}{37}\right)(2) + \left(\frac{25}{37}\right)(-1) = 0.648648648 - 0.675675675 = -0.027027027.$$

Consider the 'even chance' bet in roulette
Our expectation for a £1 bet is:

$$\left(\frac{18}{37}\right)(1) + \left(\frac{18.5}{37}\right)(-1) = -0.013513513$$

(when zero turns up, half of the stakes are lost)

KELLY'S CRITERION

Kelly is currently involved with the mathematics of gambling. He worked out the expected profit after several simulated trials.

P_w = probability of winning P_L = probability of losing.

If $P_w - P_L > 0$ that is the probability of winning is greater than the probability of losing.

If $P_w = 0.52$, $P_L = 0.48$ $P_w - P_L = 0.52 - 0.48 = 0.04$.

If the bank roll is 100 $100 \times 0.04 = 4$.

Therefore bet 4 units if you win then

$104 \times 0.04 = 4.16$ if you win

$108.16 \times 0.04 = 4.3264$ if you lose

$103.8336 \times 0.04 = 4.153344$ if you win

$107.986944 \times 0.04 = 4.31947776$ if you win

$112.3064218 \times 0.04 = 4.49225687$ if you lose

$107.8141649 - 100 = \boxed{+\ 7.8141649}$ profit.

The profit is obtained after several simulated trials.

10. COMBINATIONS. PERMUTATIONS. PROBABILITIES

PART II

SOLUTIONS 1

1. (a) (i) $^6C_2 = \dfrac{6!}{(6-2)!\ 2!} = \dfrac{6!}{4!\ 2!} = \dfrac{4!\ 5\times 6}{4!\ 1\times 2} = 15$

(ii) $^5C_2 = \dfrac{5!}{(5-2)!\ 2!} = \dfrac{5!}{3!\ 2!} = \dfrac{4\times 5}{1\times 2} = 10$

(b) (i) $^6C_3 = \dfrac{6!}{(6-3)!\ 3!} = \dfrac{6!}{3!\ 3!} = \dfrac{4\times 5\times 6}{1\times 2\times 3} = 20$

(ii) $^5C_3 = \dfrac{5!}{(5-3)!\ 3!} = \dfrac{5!}{2!\ 3!} = \dfrac{4\times 5}{1\times 2} = 10$

(c) (i) $^6C_5 = \dfrac{6!}{(6-5)!\ 5!} = \dfrac{6!}{1!\ 5!} = 6$

(ii) $^5C_5 = \dfrac{5!}{(5-5)!\ 5!} = 1.$

2. $^6C_2 = \dfrac{6!}{(6-2)!\ 2!} = \dfrac{6!}{4!\ 2!} = \dfrac{5\times 6}{1\times 2} = 15$

1,2; 1,3; 1,4; 1,5; 1,6; 2,3; 2,4; 2,5; 2,6; 3,4; 3,5; 3,6; 4,5; 4,6; 5,6.

3. $^8C_3 = \dfrac{8!}{(8-3)!\ 3!} = \dfrac{8!}{5!\ 3!} = \dfrac{6\times 7\times 8}{1\times 2\times 3} = 56$

the table becomes rather complicated.

4. (i) $\dfrac{96!}{95!} = \dfrac{96\times 95!}{95!} = 96$ (ii) $\dfrac{25!}{26!} = \dfrac{25!}{26\times 25!} = \dfrac{1}{26}$

(iii) $\dfrac{5!}{3!\ 2!} = \dfrac{3!\ 4\times 5}{3!\ 1\times 2} = 10$

(iv) $\dfrac{25!}{5!\ 20!} = \dfrac{20!\ 21 \times 22 \times 23 \times 24 \times 25}{20!\ 5!} = \dfrac{21 \times 22 \times 23 \times 24 \times 25}{1 \times 2 \times 3 \times 4 \times 5}$

$= 21 \times 22 \times 23 \times 5 = 53130.$

5. (i) $\binom{5}{2} = \dfrac{5!}{(5-2)!\ 2!} = \dfrac{5!}{3!\ 2!} = \dfrac{4 \times 5}{1 \times 2} = 10.$

(ii) $\binom{7}{3} = \dfrac{7!}{(7-3)!\ 3!} = \dfrac{7!}{4!\ 3!} = \dfrac{7 \times 6 \times 5}{1 \times 2 \times 3} = 35.$

(iii) $\binom{y}{x} = \dfrac{y!}{(y-x)!\ x!}$ provided $y \geq x$.

(iv) $\binom{n}{r} = \dfrac{n!}{(n-r)!\ r!}$ provided $n \geq r$.

6. (i) ${}^{8}C_{3} = \dfrac{8!}{(8-3)!\ 3!} = \dfrac{8!}{5!\ 3!} = \dfrac{8 \times 7 \times 6}{1 \times 2 \times 3} = 56.$

(ii) ${}^{36}C_{8} = \dfrac{36!}{(36-8)!\ 8!} = \dfrac{36!}{28!\ 8!} = 30260340.$

(iii) ${}^{n}C_{n} = \dfrac{n!}{(n-n)!\ n!} = \dfrac{n!}{0!\ n!} = \dfrac{1}{0!} = \dfrac{1}{1} = 1.$

(iv) ${}^{n}C_{1} = \dfrac{n!}{(n-1)!\ 1!} = \dfrac{n(n-1)!}{(n-1)!} = n.$

(v) ${}^{n}C_{3} = \dfrac{n!}{(n-3)!\ 3!} = \dfrac{n(n-1)(n-2)(n-3)!}{(n-3)!\ 3!}$

$= \dfrac{n(n-1)(n-2)}{1 \times 2 \times 3} = \dfrac{n(n-1)(n-2)}{6}.$

7. (i) $P_5 = 1 \times 2 \times 3 \times 4 \times 5 = 5! = 120.$

(ii) $P_3 = 1 \times 2 \times 3 = 3! = 6$

(iii) $P_r = r!$

(iv) $P_n = n!$

(v) $P_7 = 7! = 5040.$

8. (a) $\binom{15}{5}$ ways can select the administrators and $8 - 5 = 3$ lecturers are selected in $\binom{25}{3}$ ways. Therefore the total number of ways to select 5 administrators and 3 lecturers is

$$\binom{15}{5}\binom{25}{3} = \frac{15!}{10!\ 5!} \times \frac{25!}{22!\ 3!} = 6906900.$$

(b) In every group we require 6 or 7 or 8 lecturers with 6 lecturers we can form $\binom{25}{6}\binom{15}{2} = \frac{25!}{19!\ 6!} \times \frac{15!}{13!\ 2!} = 18595500$ different groups

with 7 lecturers we can form

$$\binom{25}{7}\binom{15}{1} = \frac{25!}{18!\ 7!} \times \frac{15!}{14!\ 1!} = 7210500 \text{ different groups}$$

with 8 lecturers we can form

$$\binom{25}{8}\binom{5}{0} = \frac{25!}{17!\ 8!} \times \frac{5!}{(5-0)!\ 0!} = \frac{25!}{17!\ 8!} = 1081575.$$

Therefore we can form $18595500 + 7210500 + 1081575 = 26887575$.

9. (a) $\binom{6}{3}$ ways we can select the administrators and $5 - 3 = 2$ technical staff which are selected in $\binom{4}{2}$ ways.

Therefore the number of ways in which we can select 3 administations and 2 technical staff is,

$$\binom{6}{3}\binom{4}{2} = \frac{6!}{3!\ 3!} \times \frac{4!}{2!\ 2!} = 20 \times 6 = 120 \text{ different groups.}$$

(b) In every group we require 3 or 4 technical staff.

With 3 technical staff we can form

$$\binom{4}{3}\binom{6}{2} = \frac{4!}{1!\ 3!} \times \frac{6!}{4!\ 2!} = \frac{4}{1} \times \frac{5 \times 6}{1 \times 2} = 60 \text{ different groups.}$$

With 4 technical staff we can form

$$\binom{4}{4}\binom{6}{1} = \frac{4!}{0!\ 4!} \times \frac{6!}{5!\ 1!} = 1 \times 6 = 6 \text{ different groups.}$$

Therefore we can form 120 + 6 = 126 different groups.

10. (i) $$\binom{n}{m} = \frac{n!}{(n-m)!\ m!}, \qquad \binom{n-1}{m} = \frac{(n-1)!}{(n-1-m)!\ m!},$$

$$\binom{n-1}{m-1} = \frac{(n-1)!}{(n-1-m+1)!\,(m-1)!} = \frac{(n-1)!}{(n-m)!\ (m-1)!}$$

$$\binom{n-1}{m} + \binom{n-1}{m-1} = \frac{(n-1)!}{(n-1-m)!\ m!} + \frac{(n-1)!}{(n-m)!\ (m-1)!}$$

$$= \frac{n(n-1)!}{n(n-1-m)!\ m!} + \frac{n(n-1)!}{n(n-m)!\ (m-1)!}$$

$$= \frac{n!}{n(n-1-m)!\,(m-1)!\,m} + \frac{n!\,m}{n(n-m)!\,(m-1)!\,m}$$

$$= \frac{n!}{m!}\left[\frac{1}{n(n-1-m)!} + \frac{m}{n(n-m)!}\right]$$

$$= \frac{n!}{m!}\left[\frac{n-m}{n(n-m)\ (n-m-1)!} + \frac{m}{n(n-m)!}\right]$$

since $(n - m)! = (n - m)(n - m - 1)!$

$$= \frac{n!}{m!}\left[\frac{n - m}{n(n - m)!} + \frac{m}{n(n - m)!}\right]$$

$$= \frac{n!}{m!}\left[\frac{n}{n(n - m)!} - \frac{m}{n(n - m)!} + \frac{m}{n(n - m)!}\right]$$

$$= \frac{n!}{m!\,(n - m)!}$$

(ii) $\binom{n}{m} = \binom{n-2}{m} + 2\binom{n-2}{m-1} + \binom{n-2}{m-2}$

$$= \frac{(n-2)!}{(n-m-2)!\,m!} + \frac{2(n-2)!}{(n-m-1)!\,(m-1)!} + \frac{(n-2)!}{(n-m)!\,(m-2)!}$$

$$= \frac{(n-2)!\,(n-m-1)}{(n-m-2)!\,m!\,(n-m-1)} + \frac{2(n-2)!\,m}{(n-m-1)!\,m(m-1)!} +$$

$$\frac{(n-2)!\,(m-1)}{(n-m)!\,(m-2)!\,(m-1)}$$

$$= \frac{(n-2)!\,(n-m-1)}{(n-m-1)!\,m!} + \frac{2m\,(n-2)!}{(n-m-1)!\,m!} +$$

$$\frac{(n-2)!\,(m-1)\,m}{(n-m)!\,(m-1)!\,m}$$

$$= \frac{(n-2)!\,(n-m-1)\,(n-m)}{(n-m-1)!\,(n-m)\,m!} + \frac{2m(n-2)!\,(n-m)}{(n-m-1)!\,(n-m)\,m!} +$$

$$\frac{(n-2)!\,m\,(m-1)}{(n-m)!\,m!}$$

$$= \frac{(n-2)!\,(n-m-1)\,(n-m)}{(n-m)!\;m!} + \frac{2m\,(n-2)!\,(n-m)}{(n-m)!\;m!} +$$

$$\frac{(n-2)!\;m\,(m-1)}{(n-m)!\;m!}$$

$$= \frac{(n-2)!(n-m-1)(n-m) + 2m\,(n-2)!(n-m) + (n-2)!\,m(m-1)}{(n-m)!\,m!}$$

$$= \frac{(n-2)!\,\left[n^2 - nm - n - nm + m^2 + m + 2nm - 2m^2 + m^2 - m\right]}{(n-m)!\;m!}$$

$$= \frac{(n-2)!\,(n^2-n)}{(n-m)!\;m!} = \frac{(n-2)!\;n(n-1)}{(n-m)!\;m!} = \frac{n!}{(n-m)!\;m!} = \binom{n}{m}$$

$$\boxed{\binom{n}{m} = \binom{n-2}{m} + 2\binom{n-2}{m-1} + \binom{n-2}{m-2}}$$

Alternatively since $\binom{n}{m} = \binom{n-1}{m} + \binom{n-1}{m-1}$ replace n by $n-1$

then $\binom{n-1}{m} = \binom{n-2}{m} + \binom{n-2}{m-1}$ and replace m by $m-1$

then $\binom{n-1}{m-1} = \binom{n-2}{m-1} + \binom{n-2}{m-2}$

$$\binom{n-1}{m} + \binom{n-1}{m-1} = \binom{n-2}{m} + \binom{n-2}{m-1} + \binom{n-2}{m-1} + \binom{n-2}{m-2}$$

$$\therefore \quad \boxed{\binom{n}{m} = \binom{n-2}{m} + 2\binom{n-2}{m-1} + \binom{n-2}{m-2}}\,.$$

11. (i) $\binom{n}{5} = \frac{3}{4}\binom{n}{6} \Rightarrow \frac{n!}{(n-5)!\,5!} = \frac{3}{4}\,\frac{n!}{(n-6)!\,6!}$

$$4 \times 6!\,(n-6)! = 3 \times 5!\,(n-5)!$$

$$4 \times 6!\,(n-6)! = 3 \times 5!\,(n-6)!\,(n-5) \qquad \text{NB } 6! = 6 \times 5!$$

$$4 \times 6 = 3\,(n-5)$$

$$8 = n - 5$$

$$\boxed{n = 13}$$

(ii) $\binom{n}{26} = \frac{1}{2}\binom{n}{25} \Rightarrow \frac{n!}{(n-26)!\,26!} = \frac{1}{2}\,\frac{n!}{(n-25)!\,25!} \Rightarrow$

$$2 \times 25!\,(n-25)! = (n-26)!\,26!$$

$$2 \times (n-25) = 26 \Rightarrow n = 13 + 25 = 38 \Rightarrow \boxed{n = 38}$$

(iii) $\binom{n}{7} = \frac{1}{8}\binom{n}{8} \Rightarrow \frac{n!}{(n-7)!\,7!} = \frac{1}{8}\,\frac{n!}{(n-8)!\,8!} \Rightarrow$

$$8\,(n-8)!\,8! = (n-7)!\,7!$$

$$8 \times 8 = n - 7 \Rightarrow \boxed{n = 71}$$

SOLUTIONS 2

1. $x\,2! = 5!$, $x = \frac{5!}{2!}$, x is the number of arrangements, $2!$ is the two identical 0, and $5!$ the total number of letters

$$x = \frac{5!}{2!} = \frac{1 \times 2 \times 3 \times 4 \times 5}{1 \times 2}$$

$x = 60.$

2. $x\,2!\,2!\,2!\,2! = 11!$

$2R$, $2A$, $2E$, $2N$

$$x = \frac{11!}{2!\,2!\,2!\,2!} = 3 \times 4 \times 5 \times 6 \times 7 \times 9 \times 10 \times 11 = 2494800.$$

3. $x = 9!$

no identical letters.

$x = 9! = 362880.$

4. There are $(3B)$ and $(4C)$ and the total number of letters is 10.

$x\,3!\,4! = 10!$

$$x = \frac{10!}{3!\,4!} = 25200.$$

5. (i) $${}^5P_3 = \frac{5!}{(5-3)!} = \frac{5!}{2!} = 3 \times 4 \times 5 = 60.$$

(ii) $${}^8P_3 = \frac{8!}{(8-3)!} = \frac{8!}{5!} = 6 \times 7 \times 8 = 336.$$

(iii) $${}^nP_r = \frac{n!}{(n-r)!}.$$

6. $${}^nP_r = \frac{n!}{(n-r)!} \text{ and } {}^nC_r = \frac{n!}{(n-r)!\,r!}$$

$$\boxed{{}^nC_r = {}^nP_r \div r!}$$

$$\boxed{{}^nP_r = r!\,{}^nC_r}$$

7. $${}^nP_r = \frac{n!}{(n-r)!} = 40320$$

$$\binom{n}{r} = {}^nC_r = 56 = \frac{n!}{(n-r)!\, r!}$$

$$r! \; \frac{40320}{56} = 720 = 6!$$

$$\boxed{r = 6}$$

8. $${}^5P_3 = \frac{5!}{(5-3)!} = \frac{5!}{2!} = 3 \times 4 \times 5 = 60.$$

9. $${}^7P_4 = \frac{7!}{(7-4)!} = \frac{7!}{3!} = 4 \times 5 \times 6 \times 7 = 840.$$

10. The American can seat in 7! ways.
The German can seat in 5! ways.
The British can seat in 4! ways.
The three nationalities can seat in 3! ways. Therefore the number of ways that they can seat adjacent to each other are 3! 7! 5! 4! = 87091200.

11. The first three digits of the number can be chosen in ${}^5P_3$ ways (since there are five odd numbers and we require any three of them). The last two digits which are to be even can be chosen in ${}^4P_2$ ways (since there are four even numbers and we require any two of them).

Therefore we can form ${}^5P_3 \,.\, {}^4P_3 = \frac{5!}{3!}\frac{4!}{2!} = 4 \,.\, 5 \,.\, 3 \,.\, 4 = 240.$

12. (a) We can form 6! words, that is, 720 words.

(b) One such word shall be *VCVCVC* (V = vowel, C = consonant). The three vowels can be placed in 3! ways and the three consonants can also be placed in 3! ways. Therefore the number of ways we can form words of the form *VCVCVC* are 3! 3! = 6 × 6 = 36. Also there are 36 words of the form *CVCVCV* and therefore we have 72 such words with *VC* or *CV* alternating.

13. (a) (i) There are six different letters and the total number of words including the word *CEYLON* is the permutation of the six letters, 6! = 720.

(ii) There are six different letters and the total number of words including the word *FORMAT* is the permutation of the six letters, is $6! = 720$.

(b) (i) For the first letter we can select one out of three consonants and for the last letter we can also select one out of the three vowels and the number of the middle letters are the permutation of 4 letters, $4! = 24$. Therefore the number of words are $3 \times 3 \times 24 = 216$.

(ii) For the first letter we can select one out of the four consonants and for the last letter we can select one of the two vowels, the number of the middle four letters is the permutation of the four letters, that is $4! = 24$. Therefore the number of words required are $4 \times 2 \times 24 = 192$ words.

14. $$\sum_{r=0}^{n} 2^r \binom{n}{r} = 2^0 \binom{n}{0} + 2^1 \binom{n}{1} + 2^2 \binom{n}{2} + \ldots + 2^n \binom{n}{n}$$

$$= 2^0 \frac{n!}{0!\, n!} + 2^1 \frac{n!}{(n-1)!\, 1!} + 2^2 \frac{n!}{(n-2)!\, 2!} + \ldots + 2^n \frac{n!}{n!\, 0!}$$

$$= 1 + 2^1 \frac{(n-1)!\, n}{(n-1)!\, 1!} + 2^2 \frac{(n-2)!\,(n-1)\, n}{(n-2)!\, 2!} + \ldots + 2^n$$

$$= 1 + 2\frac{n}{1!} + 2^2 \frac{n(n-1)}{2!} + \ldots + 2^n = (1+2)^n = 3^n$$

The expansion of

$$(1+2)^n = 1 + 2n + \frac{n(n-1)}{1 \times 2} 2^2 + \frac{n(n-1)(n-2)}{1 \times 2 \times 3} 2^3 + \ldots + 2^n$$

therefore $$\boxed{3^n = \sum_{r=0}^{n} 2^n \binom{n}{r}}.$$

SOLUTIONS 3, 4 AND 5

1. The probability of throwing one head with one throw of a coin is $\frac{1}{2}$, the probability of throwing two heads with two throws of a coin is $\frac{1}{2} \times \frac{1}{2}$, with three throws, $P = \frac{1}{2} \times \frac{1}{2} \times \frac{1}{2}$ and of course of throwing 5 consecutive heads with five throws of a coin is $\frac{1}{2} \times \frac{1}{2} \times \frac{1}{2} \times \frac{1}{2} \times \frac{1}{2} = \frac{1}{32}$.

2. $P = \frac{1}{2} \times \frac{1}{2} \times \frac{1}{2} = \frac{1}{8}$.

3. $P = \frac{1}{13} \times \frac{1}{4} = \frac{1}{52}$.

   The king of spades is one card out of a pack of 52, so the probability of drawing it is $\frac{1}{52}$.

4. The probability that the first ball drawn is red is $\frac{3}{12}$ since there are 3 reds and the total number of balls is 12, we have now left $2R$, $4Y$, $5B$, a total of 11 balls, the probability of drawing the second red ball is $\frac{2}{11}$, the number of reds is 2 and the total remaining balls is 11.

   There are now 10 balls left, the probability of drawing a blue ball is $\frac{5}{10}$.

   Therefore, the probability of drawing two red balls and one blue ball is

   $= \frac{3!}{2!\ 1!} \times \frac{3}{12} \times \frac{2}{11} \times \frac{5}{10} = \frac{9}{132}$. since there are $\frac{3!}{2!\ 1!} = 3$ ways of drawing two red and one blue ball (*RRB, RBR, BRR*).

5. The probability of drawing a red disc is $\frac{5}{12}$, we have now a total of 11 discs the probability of drawing a yellow disc is $\frac{4}{11}$, we have now a total of 10 discs, the probability of drawing a blue disc is $\frac{3}{10}$.

The probability of drawing 1 red, 1 yellow and 1 blue disc is

$$= \frac{3!}{1!\ 1!\ 1!} \times \frac{5}{12} \times \frac{4}{11} \times \frac{3}{10} = \frac{5}{110}.$$

Since there are $3! = 6$ ways of drawing 1 red, 1 yellow and 1 blue.
(*RYB, RBY, YRB, YBR, BRY, BYR*)

6. $\frac{1}{3}$ is the probability of drawing a blue ball, now there are two balls left, the probability of drawing a yellow ball is $\frac{1}{2}$ and one ball is now left, the red one, the probability of picking this is certainly 1. The probability of drawing 1 Blue, 1 Yellow and 1 Red in that order is $\frac{1}{3} \times \frac{1}{2} \times \frac{1}{1} = \frac{1}{6}$.

7. (i) $^{17}C_7 = \frac{17!}{(17 - 7)!\ 7!} = \frac{17!}{10!\ 7!}$

$$P = \frac{17!}{10!\ 7!} \times \left(\frac{1}{2}\right)^{17}$$

(ii) $^{17}C_{10} = \frac{17!}{(17 - 10)!\ 10!} = \frac{17!}{7!\ 10!}$

$$P = \frac{17!}{10!\ 7!} \times \left(\frac{1}{2}\right)^{17}$$

8. $^8C_6 = \dfrac{8!}{(8-6)!\,6!} = \dfrac{7 \times 8}{1 \times 2} = 28$

$$P = 28 \times \left(\frac{1}{2}\right)^8 = \frac{7}{64}.$$

9. (i) $P(A \cap B) = P(A \text{ and } B) = P(A) \times P(B)$

$= 0.6 \times 0.7 = 0.42$

(ii) $P(A \cup B) = P(A \text{ or } B \text{ or both}) = P(A) + P(B) - P(A \cap B)$

$= 0.6 + 0.7 - 0.42 = 1.3 - 0.42 = 0.88.$

10. (i) $P(A \cap B) = P(A) \times P(B) = 0.1 \times 0.2 = 0.02$

(ii) $P(A \cup B) = P(A) + P(B) - P(A \cap B)$

$= 0.1 + 0.2 - 0.02 = 0.3 - 0.02 = 0.28.$

11. $P(A \cap B) = 0.1,\ P(A \cup B) = 0.8$

$P(A \cap B) = P(A) \times P(B) = 0.1 \quad \dots (1)$

$P(A \cup B) = P(A) + P(B) - P(A \cap B)$

$P(A) + P(B) = P(A \cup B) + P(A \cap B)$

$= 0.8 + 0.1 = 0.9.$

The sum of probabilities is 0.9 the product of probabilities is 0.1.
If x_1 and x_2 are the probabilities, the quadratic equation is given by
$x^2 - (\text{sum of } x_1\ x_2)\ x + (\text{Product of } x_1\ x_2) = 0$

$x^2 - 0.9x + 0.1 = 0$

Note: If $ax^2 - bx + c = 0$ then $x = \dfrac{-b \pm \sqrt{b^2 - 4ac}}{2a}$, if $a > 0$

$$x = \frac{0.9 \pm \sqrt{0.9^2 - 4 \times 0.1}}{2} = \frac{0.9 \pm \sqrt{0.81 - 0.4}}{2}$$

$$x = \frac{0.9 + \sqrt{0.41}}{2} = 0.77 \quad \text{or} \quad x = \frac{0.9 - \sqrt{0.41}}{2} = 0.13$$

$P(A) = 0.77016$ or $P(B) = 0.12984$ or vice-versa.

12. $P(A \cap B) = P(A) \times P(B) = 0.07$

$P(A \cup B) = P(A) + P(B) - P(A \cap B)$

$P(A) + P(B) = P(A \cup B) + P(A \cap B) = 0.7 + 0.07 = 0.77.$

The sum of probabilities is 0.77 the product of probabilities is 0.07

$x^2 - 0.77x + 0.07 = 0$

$$x = \frac{0.77 \pm \sqrt{0.77^2 - 4 \times 0.07}}{2} = \frac{0.77 \pm 0.55937}{2}$$

$P(A) = 0.6647$ or $P(B) = 0.1053$ or vice-versa.

13. (a) The total outcomes is nine since the numbers 1, 2 and 6 are likely to occur twice, the probability of throwing a 1 is $\frac{2}{9}$.

(b) The probability of throwing a 6 is also $\frac{2}{9}$.

(c) The probability of throwing a three is $\frac{1}{9}$.

14. (a) The total outcomes is 10, the probability of throwing a 3 is $\frac{3}{10}$.

(b) Given that the throw is greater than 3; we disregard 1, 2 and 3, which give five outcomes, then the probability of throwing a 4 is $\frac{3}{5}$.

15. (i) $\dfrac{4}{32} = \dfrac{1}{8}$

(ii) $\dfrac{4}{32} \times \dfrac{3}{31} = 0.01210$

(iii) $\dfrac{4}{32} \times \dfrac{3}{31} \times \dfrac{2}{30} = 0.0008065$

(iv) $\dfrac{4}{32} \times \dfrac{3}{31} \times \dfrac{2}{30} \times \dfrac{1}{29} = 0.00002781.$

16. The probability of drawing the first Q is $\dfrac{4}{32}$, the second Q is $\dfrac{3}{31}$, the third Q is $\dfrac{2}{30}$, the first J is $\dfrac{4}{29}$, the second J is $\dfrac{3}{28}$. The probability of drawing

$3Q$ and $2J$ = $\dfrac{5!}{3!\ 2!} \times \dfrac{4}{32} \times \dfrac{3}{31} \times \dfrac{2}{30} \times \dfrac{4}{29} \times \dfrac{3}{28} = 1.1918 \times 10^{-4}$,

since there are $\dfrac{5!}{3!\ 2!} = 10$ ways of selecting $3Q$ and $5J$.

17. (a) The probability of drawing the first ace is $\dfrac{4}{52} = \dfrac{1}{13}$, the probability of drawing the second ace is $\dfrac{3}{51}$, the probability of drawing the third ace is $\dfrac{2}{50}$ and the probability of drawing the fourth ace is $\dfrac{1}{49}$, the probability of drawing 1 king is $\dfrac{4}{48}$.

There are $\dfrac{5!}{4!\ 1!} = 5$ ways of selecting 4 aces and 1 king.

The probability of drawing 4 aces and a king is

$$\frac{5!}{4!\ 1!} \times \frac{1}{13} \times \frac{3}{51} \times \frac{2}{50} \times \frac{1}{49} \times \frac{4}{48} = 1.54 \times 10^{-6}$$

(b) $\dfrac{5!}{3!\ 2!} \times \dfrac{1}{13} \times \dfrac{3}{51} \times \dfrac{2}{50} \times \dfrac{4}{49} \times \dfrac{3}{48} = 9.23 \times 10^{-6}.$

10. COMBINATIONS. PERMUTATIONS. PROBABILITIES

MISCELLANEOUS EXERCISES WITH SOLUTIONS

1. (i) The events A and B are such that

$$P(A) = 0.4, \; P(B) = 0.45, \; P(A \cup B) = 0.68.$$

Show that the events A and B are neither mutually exclusive nor independent.

(ii) A bag contains 12 red balls, 8 blue balls and 4 white balls. Three balls are taken from the bag at random and without replacement. Find the probability that all three balls are of the same colour.

Find also the probability that all three balls are of different colours.

June 1982 U.L.

SOLUTION 1

(i) $P(A) = 0.4 \qquad P(B) = 0.45 \qquad P(A \cup B) = 0.68$

NOT MUTUALLY EXCLUSIVE IF

$$P(A \cup B) = P(A) + P(B) - P(A \cap B)$$

$$0.68 = 0.4 + 0.45 - P(A \cap B)$$

$$P(A \cap B) = 0.17$$

NOT INDEPENDENT AS $P(A \cap B) \neq P(A)\,P(B)$

$$0.17 \neq 0.4 \times 0.45 = 0.18$$

(ii) Bag 12 Red balls
8 Blue balls
4 White balls

P (3 balls are same colour) $= P(RRR \text{ or } BBB \text{ or } WWW)$

$= P(R)P(R)P(R) + P(B)P(B)P(B) + P(W)P(W)P(W)$

$$= \frac{12}{24}\frac{11}{23}\frac{10}{22} + \frac{8}{24}\frac{7}{23}\frac{6}{22} + \frac{4}{24}\frac{3}{23}\frac{2}{22}$$

$$= \frac{1320 + 336 + 24}{24 \times 23 \times 22} = \frac{1680}{12144} = \frac{35}{253}$$

P (ALL 3 balls are of different colours) $= \boldsymbol{P}\ (\boldsymbol{RWB})$

$$= \frac{3!}{1!\ 1!\ 1!} \boldsymbol{P}\ (\boldsymbol{R})\ \boldsymbol{P}\ (\boldsymbol{W})\ \boldsymbol{P}\ (\boldsymbol{B})$$

$$= 6 \cdot \frac{12}{24} \cdot \frac{4}{23} \cdot \frac{8}{22} = \frac{48}{253}.$$

2. A trial consists of rolling a red die and a blue die. The score T resulting from the trials is defined as the sum of the numbers showing when the numbers on the red and the blue dice are the same, but as the product of these numbers when they are different. Find (a) $P\ (T = 6)$, (b) $P\ (T = 8)$, (c) $P\ (T = 12)$.

If the scores from two trials are to be added together, find the probability of getting more than 45.

SPECIAL 1982 U.L.

SOLUTION 2

Dice **Red**	**Blue**	T (score)
1	1	2
2	1	2
1	2	2
2	2	4
1	3	3
3	1	3
3	2	6
2	3	6
3	3	6

(a) $P\ (T = 6) = \frac{5}{36}$

(b) $P\ (T = 8) = \frac{3}{36}$

(c) $P\ (T = 12) = \frac{5}{36}$

4	1	4
1	4	4
4	2	8
2	4	8
4	3	12
3	4	12
4	4	8
5	1	5
1	5	5
5	2	10
2	5	10
5	3	15
3	5	15
5	4	20
4	5	20
5	5	10
6	1	6
1	6	6
6	2	12
2	6	12
6	3	18
3	6	18
6	4	24
4	6	24
6	5	30
5	6	30
6	6	12

If scores are 'added' from 2 trials

$$P(>45) = P(30 \text{ and } 30 \text{ or } 30 \text{ and } 24 \text{ or } 30 \text{ and } 18 \text{ or } 24 \text{ and } 24 \text{ or } 20 \text{ and } 30)$$

$$= P(30) \times P(30) + \frac{2!}{1!\,1!} P(30)\, P(24) + \frac{2!}{1!\,1!} P(30)\, P(18) + P(24)\, P(24) + \frac{2!}{1!\,1!} \times P(30) \times P(20)$$

$$= \frac{2}{36} \times \frac{2}{36} + 2\left(\frac{2}{36}\right) \times \left(\frac{2}{36}\right) + 2\left(\frac{2}{36}\right)\left(\frac{2}{36}\right) + \left(\frac{2}{36}\right)\left(\frac{2}{36}\right) + 2 \times \frac{2}{36} \times \frac{2}{36}$$

$$= \frac{4}{36^2} + \frac{8}{36^2} + \frac{8}{36^2} + \frac{4}{36^2} + \frac{8}{36^2}$$

$$= \frac{32}{36 \times 36}$$

$$= \frac{2}{9} \times \frac{1}{9} = \boxed{\frac{2}{81}}.$$

3. (i) Two cards are drawn without replacement from ten cards which are numbered from 1 to 10. Find the probability that

 (a) the numbers on both cards are even,

 (b) the number on one card is odd and the number on the other card is even,

 (c) the sum of the numbers on the two cards exceeds 4.

 (ii) Events A and C are independent. Probabilities relating to events A, B and C are as follows: **Jan. 1983 U.L.**

$$P(A) = \frac{1}{5},\ P(B) = \frac{1}{6},\ P(A \cap C) = \frac{1}{20},\ P(B \cup C) = \frac{3}{8}.$$

Evaluate $P(C)$ and hence show that events B and C are independent.

Jan. 1983 U.L.

SOLUTION 3

(i) 10 cards 1, 2, 3, 4, 5, 6, 7, 8, 9, 10 Remove 2

(a) P (BOTH CARDS ARE EVEN) $= P(E \text{ and } E) = P(E)\,P(E)$

$$= \frac{5}{10} \times \frac{4}{9} = \boxed{\frac{2}{9}}$$

(b) P (ONE CARD IS ODD AND ONE CARD IS EVEN)

$$= \frac{2!}{1!\ 1!} P(O) \times P(E) = 2 \times \frac{5}{10} \times \frac{5}{9} = \boxed{\frac{5}{9}}$$

(c) P (SUM OF 2 NUMBERS EXCEED 4)

$= P$ (1 , 4 or 1, 5 or 1, 6 or 1, 7 or 1, 8 or 1, 9 or 1, 10 or 2, 3, or 2, 4... etc.)

$= 1 - P\{1, 2 \text{ or } 1, 3\}$

$$= 1 - \left(\frac{2!}{1!\ 1!} P(1)\,P(2) + \frac{2!}{1!\ 1!} P(1)\,P(3)\right)$$

$$= 1 - \left(2 \times \frac{1}{10} \times \frac{1}{9} + 2 \times \frac{1}{10} \times \frac{1}{9}\right) = 1 - \frac{4}{90} = \frac{86}{90} = \frac{43}{45}$$

(ii) AC independent

$$P(A) = \frac{1}{5} \qquad P(B) = \frac{1}{6} \qquad P(A \cap C) = \frac{1}{20} \qquad P(B \cup C) = \frac{3}{8}$$

Find $P(C)$

If independent $\quad P(A \cap C) = P(A)\,P(C)$

$$\frac{1}{20} = \frac{1}{5} P(C)$$

$$\boxed{P(C) = \frac{1}{4}}$$

If B, C are independent

$$P(B \cap C) = P(B)\,P(C)$$

but $P(B \cup C) = P(B) + P(C) - P(B \cap C)$

$$\frac{3}{8} = \frac{1}{6} + \frac{1}{4} - P(B \cap C)$$

$$P(B \cap C) = \frac{1}{6} + \frac{1}{4} - \frac{3}{8} = \frac{4 + 6 - 9}{24} = \frac{1}{24}$$

$$\boxed{P(B \cap C) = \frac{1}{24}}.$$

However

As $P(B \cap C) = P(B)\,P(C)$

$\frac{1}{24} = \frac{1}{4} \cdot \frac{1}{6}$ $\therefore$ B, C are independent.

4. For events A and B, express $P(A \cup B)$ in terms of $P(A)$ and $P(B)$ only, when (a) A and B are mutually exclusive,
(b) A and B are independent.

A bag contains just 10 balls, of which 5 are red and 5 are black. One ball is drawn at random from the bag and replaced; a second ball is drawn at random and replaced and then a third ball is drawn at random. By means of a tree diagram, or otherwise, show that the probability of drawing 2 black balls and one red ball is $\frac{3}{8}$.

Event A is that the three balls drawn include at least one red ball and at least one black ball. Event B is that the 3 balls drawn include at least 2 black balls. Find $P(A)$ and $P(B)$ and show that A and B are independent.

Given that event C is that the first 2 balls drawn are of the same colour, ascertain whether events B and C are mutually exclusive.

June 1983 U.L.

SOLUTION 4

(a) If A and B are mutually exclusive.

$$\boxed{P(A \cup B) = P(A) + P(B)}$$

(b) If A, B are independent

$$P(A \cap B) = P(A)\,P(B)$$

$$\text{but } P(A \cup B) = P(A) + P(B) - P(A \cap B)$$

$$\boxed{P(A \cup B) = P(A) + P(B) - P(A).P(B)}$$

A bag contains 10 balls {5 red and 5 black}
3 balls drawn with replacement

$$P\text{ (2 black balls and 1 red)} = \frac{3!}{2!\;1!}\,P(B)\,P(B)\,P(R)$$

$$= 3\,\frac{5}{10}\,\frac{5}{10}\,\frac{5}{10} = 3 \times \frac{1}{2} \times \frac{1}{2} \times \frac{1}{2} = \frac{3}{8}.$$

Event A 3 balls include at least 1 red and 1 black ball
Event B 3 balls include at least 2 black balls.

$$P(A) = P(RRB \text{ or } RBB) = \frac{3!}{2!\;1!}\,P(R)\,P(R)\,P(B) + \frac{3!}{2!\;1!}\,P(R)\,P(B)\,P(B)$$

$$P(A) = 3.\frac{5}{10}.\frac{5}{10}.\frac{5}{10} + 3\cdot\frac{5}{10}\cdot\frac{5}{10}\cdot\frac{5}{10} = \frac{3}{8} + \frac{3}{8} = \frac{3}{4}$$

$$P(B) = P(BBR \text{ or } BBB) = \frac{3!}{2!\;1!}\,P(B)\,P(B)\,P(R) + \frac{3!}{0!\;3!}\,P(B)\,P(B)\,P(B)$$

$$= 3.\frac{5}{10}.\frac{5}{10}.\frac{5}{10} + \frac{5}{10}.\frac{5}{10}.\frac{5}{10} = \frac{3}{8} + \frac{1}{8} = \frac{1}{2}.$$

A and B are independent if $P(A_2 \mid B_1) = P(A_1)$

$$\therefore \frac{3}{4} = \frac{3}{4}$$

If C is event that first 2 balls are of the same colour

i.e. $P(C) = P(RRB \text{ or } RRR \text{ or } BBR \text{ or } BBB)$

It is clear that B and C events are NOT mutually exclusive since events $P(BBR)$ and $P(BBB)$ occur in both B and C.

5. (i) A box contain 4 white balls and 2 black balls. Alan draws one ball. Find the probability that the ball is black.
Alan does not replace the ball and does not disclose its colour, and then Bill draws one ball. Find the probability that the ball drawn is black.

(ii) A child has two full sets of 26 alphabet blocks. He draws 4 blocks from each set. Obtain, but do not evaluate, expressions for the probabilities that from the 8 blocks he draws he can form

(a) the four letter word *ABBA*

(b) the seven letter word *GLENELG*.

Jan. 84 U.L.

SOLUTION 5

(i) Box = $4W$ balls and $2B$ balls

$$P\text{ (Alan black)} = \frac{2}{6} = \frac{1}{3}$$

$$P\text{ (Bill draws a black)} = P\text{ (Alan } W \text{ and Bill } B \text{ or Allan } B \text{ and Bill } B)$$

$$= P(W)\,P(B) + P(B)\,P(B)$$

$$= \frac{4}{6} \times \frac{2}{5} + \frac{2}{6} \times \frac{1}{5}$$

$$= \boxed{\frac{1}{3}}$$

(ii) 2 sets of 26 alphabet blocks.

4 drawn from each set

P (8 blocks from the word *ABBA*)

A represent the letter A

B represent the letter B

N_{AB} represent all letters except A and B

$$= P\left(N_{AB}, N_{AB}, A, B \text{ and } B, A, N_{AB}, N_{AB}\right)$$

$$= \frac{4!}{2!\ 1!\ 1!} \times \frac{24}{26} \times \frac{23}{25} \times \frac{1}{24} \times \frac{1}{23} \times \frac{4!}{2!\ 1!\ 1!} \times \frac{1}{26} \times \frac{1}{25} \times \frac{24}{24} \times \frac{23}{23}$$

$$= \left(\frac{12}{26 \times 25}\right)^2 = \left(\frac{6}{13 \times 25}\right)^2$$

P (seven letter word *GLENELG*)

$= P$ (*GLENELG-* or *-GLENELG*

Block 1 2 3 4	Block 1 2 3 4
GLEN	*ELG-*
-GLE	*NELG*

$$= 4! \times \frac{1}{26} \times \frac{1}{25} \times \frac{1}{24} \times \frac{1}{23} \times 4! \times \frac{1}{26} \times \frac{1}{25} \times \frac{1}{24} \times \frac{23}{23} +$$

$$4! \times \frac{23}{26} \times \frac{1}{25} \times \frac{1}{24} \times \frac{1}{23} \times 4! \times \frac{1}{26} \times \frac{1}{25} \times \frac{1}{24} \times \frac{1}{23}$$

$$= \left(\frac{24}{26 \times 25 \times 24}\right)^2 \frac{2}{23} = \left(\frac{1}{26 \times 25}\right)^2 \frac{2}{23}.$$

6. (i) Events A and B are independent. The probability of A occurring is $\frac{1}{3}$ and the probability of B occurring is $\frac{1}{4}$. Find the probability of

(a) neither event occurring

(b) one and only one of the two events occurring.

(ii) A bag contains 10 balls, of which 3 are red, 3 are blue and 4 are white. Three balls are to be drawn one at a time, at random and without replacement, from the bag. From the probability that

(a) the first two balls drawn will be of different colours,
(b) all three balls drawn will be of the same colour,
(c) exactly 2 of the balls drawn will be of the same colour.

June 1984 U.L.

SOLUTION 6

(i) $P(A) = \frac{1}{3}$, $\quad P(A') = \frac{2}{3}$; $\quad P(B) = \frac{1}{4}$, $\quad P(B') = \frac{3}{4}$

(a) $P(A' \cap B') = P(A') \times P(B') = \frac{2}{3} \times \frac{3}{4} = \boxed{\frac{1}{2}}$

the probability of neither event occurring.

(b) P (one and only one of the two events occurring)

$= P$ (A and B' or B and A')

$= P(A) \times P(B') + P(B) \times P(A')$

$= \frac{1}{3} \times \frac{3}{4} + \frac{1}{4} \times \frac{2}{3} = \frac{1}{4} + \frac{1}{6} = \frac{3}{12} + \frac{2}{12} = \boxed{\frac{5}{12}}$

(ii) 3 Red, 3 Blue, 4 White

(a) P (TWO DIFFERENT COLOURS)

$= P$ (RB or BW or RW)

$= \frac{2!}{1!\,1!} P(R) \times P(B) + \frac{2!}{1!\,1!} P(B) \times P(W) + \frac{2!}{1!\,1!} P(R) \times P(W)$

$= 2 \times \frac{3}{10} \times \frac{3}{9} + 2 \times \frac{3}{10} \times \frac{4}{9} + 2 \times \frac{3}{10} \times \frac{4}{9}$

$= \frac{1}{5} + \frac{4}{15} + \frac{4}{15} = \frac{3}{15} + \frac{4}{15} + \frac{4}{15} = \boxed{\frac{11}{15}}$

(b) P (SAME COLOUR) $= P$ (RRR or BBB or WWW)

$= P(R) \times P(R) \times P(R) + P(B) \times P(B) \times P(B) + P(W) \times P(W) \times P(W)$

$$= \frac{3}{10} \times \frac{2}{9} \times \frac{1}{8} + \frac{3}{10} \times \frac{2}{9} \times \frac{1}{8} + \frac{4}{10} \times \frac{3}{9} \times \frac{2}{8} = \frac{6}{720} + \frac{6}{720} + \frac{24}{720}$$

$$= \boxed{\frac{1}{20}}$$

(c) P (TWO OUT OF THREE HAVE SAME COLOUR)

$$= P\left(RRR' \text{ or } BBB' \text{ or } WWW'\right)$$

$$= \frac{3!}{2!\ 1!} P(R)P(R)P(R') + \frac{3!}{2!\ 1!} P(B)P(B)P(B') + \frac{3!}{2!\ 1!} P(W)P(W)P(W')$$

$$= 3 \times \frac{3}{10} \times \frac{2}{9} \times \frac{7}{8} + 3 \times \frac{3}{10} \times \frac{2}{9} \times \frac{7}{8} + 3 \times \frac{4}{10} \times \frac{3}{9} \times \frac{6}{8}$$

$$= \frac{7}{40} + \frac{7}{40} + \frac{3}{10}$$

$$= \frac{7 + 7 + 12}{40} = \frac{26}{40}$$

$$= \boxed{\frac{13}{20}}$$

7. From past experience factory A supplies rods of which 10% are faulty. A second factory B has a better testing procedure and supplies rods of which only 5% are faulty. A retailer stocks a large number of rods of which 40% are from factory A and 60% are from factory B. Whenever a rod is sold, it is selected at random from stock. Find the probability that when a rod is sold it is faulty.
A man has to buy 4 rods from the retailer. Find, to 3 decimal places, the probability that at least one rod will be faulty.

JUNE 1985 U.L Applied

SOLUTION 7

Let A be the event that a rod comes from supplier A

B be the event that a rod comes from supplier B

F be the event that a rod is faulty.

$$P(F \mid A) = \frac{10}{100} = 0.1 \quad P(F \mid B) = \frac{5}{100} = 0.05$$

$$P(A) = \frac{40}{100} = 0.4 \quad P(B) = \frac{60}{100} = 0.6$$

$$P(\text{Faulty}) = P(A \text{ and } F \mid A \text{ or } B \text{ and } F \mid B)$$

$$= P(A) \times P(F \mid A) + P(B) \times P(F \mid B)$$

$$= 0.4 \times 0.1 + 0.6 \times 0.05 = 0.04 + 0.03 = 0.07$$

4 rods are bought

$$P(\text{at least one is FAULTY}) = P(FF'F'F' \text{ or } FFF'F' \text{ or } FFFF' \text{ or } FFFF)$$

$$= 1 - P(F'F'F'F')$$

$$= 1 - P(F')\,P(F')\,P(F')\,P(F')$$

$$= 1 - 0.93 \times 0.93 \times 0.93 \times 0.93$$

$$= 1 - 0.748 = 0.252.$$

8. In a game, each playing card in a standard pack is associated with a number, a King, Queen or Knave is associated with zero, an ace with one and each of the other cards with the number on the front of the card. Two cards are drawn from the pack at random and without replacement, and the associated numbers are recorded. When two cards are associated with the same number they are said to form a "pair". find the probability, as an exact fraction in its lowest terms, that a single draw of two cards (a) two zeros will be recorded, (b) at least one zero will be recorded, (c) the cards will form a pair, (d) the cards will form a pair and also be of the same colour.

Jan. 1985 U.L.

SOLUTION 8

(a) $J, Q, K = 0$

$$P(\text{Two zeros}) = P(z_1 \text{ and } z_2) = P(z_1) \times P(z_2)$$

$$P(z_1) = \frac{12}{52} \qquad P(z_1') = \frac{40}{52}$$

$$P(\text{Two zeros}) = P(z_1) \times P(z_2) = \frac{12}{52} \times \frac{11}{51} = \frac{11}{221}$$

(b) P (AT LEAST ONE ZERO) $= P(z, z' \text{ or } z, z)$

$$= 1 - P(z_1' z_2') = 1 - P(z_1')\,P(z_2')$$

$$= 1 - \frac{40}{52} \times \frac{39}{51} = 1 - \frac{10}{17} = \frac{7}{17}$$

(c) P (PAIR) $= P$ (1, 1 or 2, 2, or 3, 3 or 4, 4 or 5, 5 or ... or 9, 9 or 10, 10 or z, z)

$$= \frac{4}{52} \times \frac{3}{51} + \frac{4}{52} \times \frac{3}{51} + \ldots + \frac{4}{52} \times \frac{3}{51} + \frac{4}{52} \times \frac{3}{52} + \frac{12}{52} \times \frac{11}{51}$$

$$= 10\left(\frac{4}{52} \times \frac{3}{51}\right) + \frac{12}{52} \times \frac{11}{51} = \frac{252}{52 \times 51} = \frac{21}{221}$$

(d) P (PAIR AND SAME COLOUR)

i.e.	♦	♣	♥	♠	
	2	2			x
	2		2		✓
	2			2	x
		2	2		x
		2		2	✓
			2	2	x

P (same colour) $= \frac{1}{3}$

P (Pair and same colour) $= P$ (Pair) $\times P$ (same colour)

$$= \frac{21}{221} \times \frac{1}{3} = \frac{7}{221}.$$

9. Random events X and Y are mutually exclusive and events X and Z are independent. Given that

$$P(X) = \frac{1}{4},\ P(Y) = \frac{1}{5},\ P(X \cup Z) = \frac{1}{2},\ P(Y \cap Z) = \frac{1}{15},$$

find (a) $P(X \cup Y)$, (b) $P(X \cap Y')$, (c) $P(X \cap Z)$, (d) $P(Y \cap Z')$.

(ii) A bag contains 12 balls, of which 4 are red, 5 are blue and 3 are white. Three balls are to be drawn at random and without replacement. Find the probability that (a) all 3 balls will be of the same colour
(b) all 3 balls will be of the different colours.

Jan. 1986 U.L.

SOLUTION 9

(a) If $P(X \cup Y) = P(X) + P(Y)$, X and Y are mutually exclusive

If $P(X \cap Z) = P(X)\,P(Z)$, X and Z are independent

$$P(X \cup Y) = P(X) + P(Y) = \frac{1}{4} + \frac{1}{5} = \frac{5+4}{20} = \frac{9}{20}$$

(b) $P(X \cap Y') = P(X)\,P(Y') = \frac{1}{4} \times \frac{4}{5} = \frac{1}{5}$

(c) $P(X \cap Z) = P(X)\,P(Z) = \frac{1}{4}P(Z)$

$P(X \cup Z) = P(X) + P(Z) - P(X \cap Z)$

$$\frac{1}{2} = \frac{1}{4} + P(Z) - P(X \cap Z)$$

$$\frac{1}{2} = \frac{1}{4} + P(Z) - \frac{1}{4}P(Z)$$

$$\frac{1}{4} = \frac{3}{4}P(Z)$$

$$\boxed{P(Z) = \frac{1}{3}}$$

(d) $P(Y \cap Z') = P(Y)\,P(Z') = \frac{1}{5} \cdot \frac{2}{3} = \frac{2}{15}$

$P(Y \cap Z) = P(Y)\,P(Z)$ for independence

$$\frac{1}{15} = \frac{1}{5} \times \frac{1}{3}$$

(ii) *RRRR* *BBBBB* *WWW*

(a) P (same colour)

$= P(WWW \text{ or } BBB \text{ or } RRR)$

$= P(W)\,P(W)\,P(W) + P(B)\,P(B)\,P(B) + P(R)\,P(R)\,P(R)$

$$= \frac{3}{12} \times \frac{2}{11} \times \frac{1}{10} + \frac{5}{12} \times \frac{4}{11} \times \frac{3}{10} \div \frac{4}{12} \times \frac{3}{11} \times \frac{2}{10}$$

$$= \frac{6}{1320} + \frac{60}{1320} + \frac{24}{1320} = \frac{3}{44}$$

(b) P (different colours) $= \dfrac{3!}{1!\ 1!\ 1!} P(R)\, P(W)\, P(B)$

$$= 6 \times \frac{4}{12} \times \frac{3}{11} \times \frac{5}{10} = \frac{3}{11}.$$

10. The points W, X, Y and Z are marked in order on a horizontal line with the point W furthest to the left. Four beads, one red, one green, one orange and one blue are to be placed at random at W, X, Y and Z, one bead at each point. The event A is 'the red bead is to the left of both the green bead and the blue bead'.
The event B is 'the orange bead is either at X or at Y''
The event C is 'the blue bead is next to both the red bead and the green bead'.

(a) Show that $P(A) = \frac{1}{3}$, and find the values of $P(B)$ and $P(C)$.

(b) Explain why the event B and C are mutually exclusive and evaluate $P(B \cup C)$.

(c) Evaluate $P(A \cap B)$ and $P(A \cap C)$ and hence determine whether the events A and B, or A and C, or both of these pairs of events are independent **June 1986 U.L.**

SOLUTION 10

$W \quad X \quad Y \quad Z$

{Red, Green, Orange, Blue}

(a)

W	X	Y	Z
-	R	G	B
-	R	B	G
R	-	G	B
R	-	B	G
R	B	G	-
R	B	-	G
R	G	B	-
R	G	-	B

where A : Red is left of green and blue
B : Orange at X or Y.
C : Blue is next to red and green bead.

$$P(A) = \frac{8}{4 \times 3 \times 2 \times 1} = \frac{1}{3}$$

W	X	Y	Z
-	O	-	-
-	-	O	-

$$P(B) = P(O_x \text{ or } O_y) = \frac{1}{4} + \frac{1}{4} = \frac{1}{2}$$

W	X	Y	Z
G	B	R	-
R	B	G	-
-	G	B	R
-	R	B	G

$$P(C) = \frac{4}{4 \times 3 \times 2 \times 1} = \frac{1}{6}$$

(b) B and C are mutually exclusive since if the orange is at X or Y the blue cannot be next to the red and the green bead.

$$P(B \cup C) = P(B) + P(C) = \frac{1}{2} + \frac{1}{6} = \frac{4}{6} = \frac{2}{3}$$

(c) $P(A \cap B) = \frac{4}{4 \times 3 \times 2 \times 1} = \frac{1}{6}$ $\qquad$ $P(A \cap C) = \frac{2}{24} = \frac{1}{12}$

A and B are independent if $P(A \cap B) = P(A)\,P(B)$

$$\frac{1}{6} = \frac{1}{3} \times \frac{1}{2}$$

therefore A, B are independent

A and C are independent if

$$P(A \cap C) = P(A) \times P(C)$$

$$\frac{1}{12} \neq \frac{1}{3} \times \frac{1}{6}$$

$\therefore$ A, C are not independent.

11. In the manufacture of optical lenses, the final product is tested for smoothness and clarity. If both lie outside specified limits, the product is rejected. The probability that the smoothness specification will be met is 0.5 and the probability that the clarity specification will be met is 0.8. Given that the events of smoothness and clarity are independent, find the probability that, in a lens to be selected at random,

(a) both specifications will be met,
(b) at least one specification will be met,
(c) exactly one specification will be met.

Three lenses are to be chosen at random from 20 lenses, of which 8 are perfect in that they satisfy both specifications.

Find the probability that (d) none will be perfect,
(e) at least one will be perfect,
(f) exactly one will be perfect, giving your answer to two decimal places.

Jan. 1987 U.L.

SOLUTION 11

P (Smooth) $= 0.5$ $\quad\quad$ P (Clarity) $= 0.8$

Let S be the event the lens meets smoothness specification.
Let C be the event the lens meets the clarity specification

(a) $P(S \text{ and } C) = P(S) \times P(C) = 0.5 \times 0.8 = 0.4$

(b) P (at least one specification is met)

$= P(S \text{ and } C \text{ or } S \text{ and } C' \text{ or } S' \text{ and } C)$
$= 0.5 \times 0.8 + 0.5 \times 0.2 + 0.5 \times 0.8 = 0.4 + 0.1 + 0.4 = 0.9.$

(c) P (one specification is met) $= P(S \text{ and } C' \text{ or } S' \text{ and } C)$

$$= P(S)\,P(C') + P(S')\,P(C)$$

$$= 0.5 \times 0.2 + 0.5 \times 0.8 = 0.1 + 0.4 = 0.5.$$

(d) 3 lenses chosen from 20 lenses, 8 are perfect therefore 12 are not perfect
P (perfect) $= P(S \text{ and } C) = 0.4$
Let P be the event that the lens is perfect.

P (none perfect) $= P(P' \text{ and } P' \text{ and } P') = P(P') \times P(P') \times P(P')$

$$= \frac{12}{20} \times \frac{11}{19} \times \frac{10}{18} = \frac{11}{57}$$

(e) P (at least 1 perfect) $= P(P, P, P \text{ or } PPP' \text{ or } PP'P')$

$$= 1 - P(P'P'P') = 1 - \frac{11}{57} = \frac{46}{57}$$

(f) P (one perfect) $= \dfrac{3!\, P(P \text{ and } P' \text{ and } P')}{2!\ 1!}$

$$= \frac{3!}{2!\ 1!} \times \frac{8}{20} \times \frac{12}{19} \times \frac{11}{18} = \frac{44}{95} = 0.46.$$

12. In a certain team penalty shots are taken by one of three players A, B and C. For each player the probability of scoring a goal with a penalty shot is constant. The probabilities for A, B and C scoring a goal are $\frac{1}{3}$, $\frac{1}{2}$ and $\frac{1}{4}$ respectively.

During one match A takes three penalty shots. Find the probability that
(a) A scores three goals,
(b) A scores at least one goal.

During a second match each player takes one penalty shot. Find the probability that
(c) no goal is scored.
(d) exactly one goal is scored.

Given that exactly one goal is scored from a penalty in the second match, find the probability that it is scored by C.

June 1987 U.L.

SOLUTION 12

$P(A_s) = \frac{1}{3}$, $\quad P(B_s) = \frac{1}{2}$, $\quad P(C_s) = \frac{1}{4}$

(a) $P(A_s, A_s, A_s) = P(A_s)\, P(A_s)\, P(A_s) = \left(\frac{1}{3}\right)^3 = \frac{1}{27}$

(b) P (at least 1 goal by A) $= P(A_sA_sA_s \text{ or } A_sA_sA_s' \text{ or } A_sA_s'A_s')$

$$= 1 - P(A_s', A_s', A_s')$$

$$= 1 - \left(\frac{2}{3}\right)\left(\frac{2}{3}\right)\left(\frac{2}{3}\right) = 1 - \frac{8}{27} = \frac{19}{27}$$

(c) $P\text{ (No goal)} = P\left(A_s' \text{ and } B_s' \text{ and } C_s'\right) = \frac{2}{3} \times \frac{1}{2} \times \frac{3}{4} = \frac{1}{4}$

(d) $P\text{ (one goal)} = P\left(A_s \text{ and } B_s' \text{ and } C_s' \text{ or } A_s' \text{ and } B_s \text{ and } C_s' \text{ or } A_s' \text{ and } B_s' \text{ and } C_s\right)$

$$= \frac{3}{24} + \frac{6}{24} + \frac{2}{24} = \frac{11}{24}$$

$$P(C \mid \text{Goal}) \; \frac{P(C_s)}{P\text{ (one goal scored)}} = \frac{2/24}{11/24} = \frac{2}{11}.$$

13. The bag P contains ten balls of which four are white and six are red. The bag Q contains eight balls of which five are white and three and red. Two balls are to be drawn at random from the bag P and placed in Q. One ball is then to be drawn at random from the ten balls in Q and placed in P so that at the end of these stages there will be nine balls in each bag.

The event A is "The two balls drawn from P are of the same colour".
The event B is "The ball drawn from Q is white".

By using a tree diagram, or otherwise,

(a) find the value of $P(A)$,

(b) show that $P(B) = \frac{29}{50}$,

(c) show that $P(A \cap B) = \frac{13}{50}$.

Hence deduce the value of $P(A \cup B)$.

Find the conditional probability that all three balls are red, given that they are of the same colour.

Jan. 1988 U.L.

SOLUTION 13

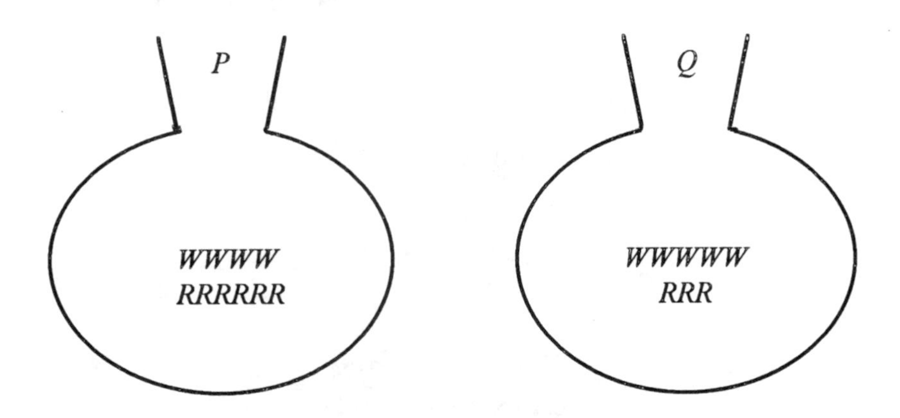

Fig. 10-M/1

2 balls are drawn from P so they could be

(1) WW i.e. Q has $7W$ $3R$
(2) WR or RW i.e. Q has $6W$ $4R$
(3) RR i.e. Q has $5W$ $5R$

(a) $P(A) = P(W \text{ and } W \text{ or } R \text{ and } R)$

$$= P(W_1)\,P(W_2) + P(R_1)\,P(R_2)$$

$$= \frac{4}{10}\cdot\frac{3}{9} + \frac{6}{10}\cdot\frac{5}{9} = \frac{12}{90} + \frac{30}{90} = \frac{42}{90} = \boxed{\frac{7}{15}}$$

$$P(\overline{A}) = 1 - P(A) = 1 - \frac{7}{15} = \frac{8}{15}$$

(b) $P(B) = P(\text{Ball from } Q \text{ is white})$

$$= P\left(W_p W_p W_Q \text{ or } W_p R_p W_Q \text{ or } R_p W_p W_Q \text{ or } R_p R_p W_Q\right)$$

$$= P(W_p)P(W_p)P(W_Q) + P(W_p)P(R_p)P(W_Q) + R(R_p)P(W_p)P(W_Q) + P(R_p)\times P(R_p)\times P(W_Q)$$

$$= \frac{4}{10}\times\frac{3}{9}\times\frac{7}{10} + \frac{4}{10}\times\frac{6}{9}\times\frac{6}{10} + \frac{6}{10}\times\frac{4}{9}\times\frac{6}{10} + \frac{6}{10}\times\frac{5}{9}\times\frac{5}{10}$$

$$= \frac{7}{75} + \frac{4}{25} + \frac{4}{25} + \frac{1}{6} = \boxed{\frac{29}{50}}.$$

(c) $P(A \cap B)$ = P (BALLS OF SAME COLOUR FROM P AND BALL FROM Q IS WHITE)

$$= P\left(W_p W_p W_Q \text{ or } R_p R_p W_Q\right)$$

$$P(W_p)P(W_p)P(W_Q) + P(R_p)P(R_p)P(W_Q) = \frac{4}{10}\times\frac{3}{9}\times\frac{7}{10}+\frac{6}{10}\times\frac{5}{9}\times\frac{5}{10}$$

(see above)

$$= \frac{7}{75} + \frac{1}{6} = \boxed{\frac{13}{50}}$$

(d) $P(A \cup P) = P(A) + P(B) - P(A \cap B)$

$$= \frac{7}{15} + \frac{29}{50} - \frac{13}{50} = \frac{59}{75}.$$

(e) P (ALL 3 BALLS ARE RED | ALL BALLS ARE ALL OF THE SAME COLOUR)

$$= \frac{P\left(R_p R_p R_Q\right)}{P\left(R_p R_p R_Q \text{ or } W_p W_p W_Q\right)} = \frac{1/6}{1/6 + 7/75} = \frac{1/6}{39/50} = \frac{25}{39}$$

$$P\left(R_p R_p R_Q\right) = P\left(R_p\right) P\left(R_p\right) P\left(R_Q\right) = \frac{6}{10}\times\frac{5}{9}\times\frac{5}{10} = \frac{1}{6}$$

$$P\left(W_p W_p W_Q\right) = \frac{7}{75}.$$

14. (i) The events A and B are such that $P(A \mid B) = \frac{7}{10}$, $P(B \mid A) = \frac{7}{15}$,

$P(A \cup B) = \frac{3}{5}$. Find the values of

(a) $P(A \cap B)$,

(b) $P(A' \cap B)$.

(ii) A hand of four cards is to be drawn without replacement and at random from a pack of fifty two playing cards. Giving your answer in each case to three significant figures, find the probabilities that this hand will contain (a) four cards of the same suit,

(b) either two aces and two Kings OR two aces and two Queens.

June 1988 U.L.

SOLUTION 14

(a) (i) $P(A \mid B) = \dfrac{P(A \cap B)}{P(B)}$ $\qquad$ $P(B \mid A) = \dfrac{P(A \cap B)}{P(A)}$

$\dfrac{7}{10} = \dfrac{P(A \cap B)}{P(B)}$ $\qquad$ $\dfrac{7}{15} = \dfrac{P(A \cap B)}{P(A)}$

$P(B) = \dfrac{10}{7} P(A \cap B) \ldots (1)$ $\qquad$ $P(A) = \dfrac{15}{7} P(A \cap B) \ldots (2)$

$$P(A \cup B) = P(A) + P(B) - P(A \cap B)$$

$$\frac{3}{5} = \frac{15}{7} P(A \cap B) + \frac{10}{7} P(A \cap B) - P(A \cap B)$$

$$\frac{3}{5} = \left(\frac{15}{7} + \frac{10}{7} - 1\right) P(A \cap B)$$

$$P(A \cap B) = \frac{3/5}{18/7} = \frac{7}{30}$$

From (1) $P(B) = \dfrac{10}{7} \cdot \dfrac{7}{30} = \dfrac{1}{3}$

From (2) $P(A) = \dfrac{15}{7} \cdot \dfrac{7}{30} = \dfrac{1}{2}$

$P(A \mid B) = \dfrac{7}{10}$ so $P(A' \mid B) = \dfrac{8}{10}$

$$P(A' \cap B) = P(A' \mid B)\, P(B) = \frac{3}{10} \times \frac{1}{3}$$

$$P(A' \cap B) = \frac{1}{10}$$

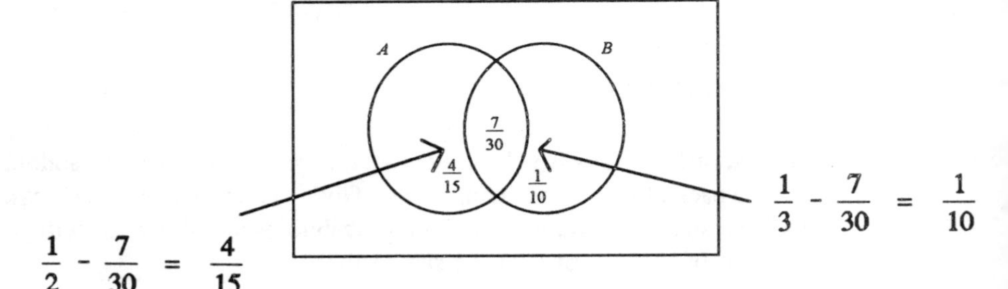

Fig. 10-M/2

(ii) 4 cards are removed.

(a) P (same suit) = P (four spades or four heart or four clubs or four diamonds)

$$P(SSSS) = \frac{13}{52} \times \frac{12}{51} \times \frac{11}{50} \times \frac{10}{49} = 0.002641$$

$$= P(CCCC) = P(DDDD) = P(HHHH).$$

Therefore, P (same suit) $= 0.002641 + 0.002641 + 0.002641 + 0.002641$

$$= 0.10564 \approx 0.0106.$$

(b) P (2 Aces and 2 Kings or 2 Aces and 2 Queens).

Let A be the event of drawing an Ace
K be the event of drawing a King
Q be the event of drawing a Queen.

$$P(A, A, K, K) = \frac{4!}{2!\ 2!} P(A_1) P(A_2) P(K_3) P(K_4)$$

$$= 6 \times \frac{4}{52} \times \frac{3}{51} \times \frac{4}{50} \times \frac{3}{49} = \frac{36}{270725} = 0.000133.$$

$$P(A, A, Q, Q) = \frac{4!}{2!\ 2!} P(A_1) P(A_2) P(Q_3) P(Q_4)$$

$$= 6 \times \frac{4}{52} \times \frac{3}{51} \times \frac{4}{50} \times \frac{3}{49} = \frac{36}{270725} = 0.000133.$$

So $P((2A \cap 2K) \cup (2A \cap 2Q)) = 0.000266.$

15. (i) Events A and B are such that $P(A) = \frac{2}{5}$, $P(B) = \frac{1}{4}$, and

$P(A \cup B) = \frac{11}{20}$. Determine whether or not the events A and B

are (a) independent,

(b) mutually exclusive.

A third event C is such that $P(A \cup C) = \frac{7}{10}$, $P(B \cup C) = \frac{3}{4}$ and $P(A \cap C) = 2P(B \cap C)$.

(c) Find, $P(C)$ and determine whether or not the events B and C are independent.

(ii) A biased die is constructed so that each of the numbers 3 and 4 is twice as likely to occur as each of the numbers 1, 2, 5 and 6. Find
(a) the probability of throwing a 4,
(b) the probability of throwing a 4, given that the throw is greater than 2.

Two such dice are thrown.
(c) Find the probability that the sum of the numbers thrown is 7.

Jan. 1989 U.L.

SOLUTION 15

(i) $P(A \cup B) = P(A) + P(B) - P(A \cap B)$

$$\frac{11}{20} = \frac{2}{5} + \frac{1}{4} - P(A \cap B)$$

$$P(A \cap B) = \frac{2}{5} + \frac{1}{4} - \frac{11}{20} = \frac{8+5}{20} - \frac{11}{20} = \frac{2}{10} = \frac{1}{10}.$$

Therefore, A and B are not mutually exclusive.

A, B are independent if $P(A \cap B) = P(A) \times P(B)$

$$P(A \cap B) = P(A) \times P(B)$$

$$\frac{1}{10} = \frac{2}{5} \times \frac{1}{4}$$

$$\frac{1}{10} = \frac{1}{10}.$$

Therefore, A and B are independent

$$P(A \cup C) = P(A) + P(C) - P(A \cap C)$$

$$\frac{7}{10} = \frac{2}{5} + P(C) - 2P(B \cap C)$$

$$\frac{7}{10} - \frac{4}{10} = P(C) - 2P(B \cap C)$$

$$\boxed{\frac{3}{10} = P(C) - 2P(B \cap C) \quad \dots (1)}$$

$$P(B \cup C) = P(B) + P(C) - P(B \cap C)$$

$$\frac{3}{4} = \frac{1}{4} + P(C) - P(B \cap C)$$

$$\frac{3}{4} - \frac{1}{4} = \boxed{\frac{1}{2} = P(C) - P(B \cap C) \quad \dots (2)}$$

Subtracting (1) from (2)

$$\frac{5}{10} - \frac{3}{10} = \boxed{P(B \cap C) = \frac{1}{5}} \quad \dots (3)$$

Substituting for $P(B \cap C)$ in (1)

$$\frac{3}{10} = P(C) - \frac{2}{5} \Rightarrow P(C) = \frac{3}{10} + \frac{4}{10} = \frac{7}{10}$$

$$\boxed{P(C) = \frac{7}{10}}$$

B, C are independent if $P(B \cap C) = P(B) \times P(C)$

$$\frac{1}{5} \neq \frac{1}{4} \times \frac{7}{10} = \frac{7}{40}.$$

Therefore, B, C are not independent.

(ii)

X	1	2	3	4	5	6
$P(X)$	$\frac{1}{8}$	$\frac{1}{8}$	$\frac{2}{8}$	$\frac{2}{8}$	$\frac{1}{8}$	$\frac{1}{8}$

Numerator totals 8 therefore the denominator must be 8 so that the probabilities sum up to 1.

Table shows the chances of scoring 3 and 4 are twice as large as other scores

(a) $P(4) = \frac{2}{8} = \frac{1}{4}$

(b) $P(4 \mid > 2) = \dfrac{2/8}{\frac{2}{8} + \frac{2}{8} + \frac{1}{8} + \frac{1}{8}} = \frac{1}{3}$.

Two dice are thrown.

(c) P (sum of 7) $= P$ (3, 4 or 2, 5 or 1, 6)

$$P(3\,,4) = \frac{2!}{1!\ 1!} P(3)\,P(4) = 2\left(\frac{2}{8}\right)\left(\frac{2}{8}\right) = \frac{8}{64} = \frac{1}{8}$$

$$P(2\,,5) = \frac{2!}{1!\ 1!} P(2)\,P(5) = 2\left(\frac{1}{8}\right)\left(\frac{1}{8}\right) = \frac{2}{64}$$

$$P(1\,,6) = \frac{2!}{1!\ 1!} P(1)\,P(6) = 2\left(\frac{1}{8}\right)\left(\frac{1}{8}\right) = \frac{2}{64}$$

$$P(\text{sum of } 7) = \frac{8}{64} + \frac{2}{64} + \frac{2}{64} = \frac{12}{64} = \frac{3}{16}.$$

16. The events A and B are such that $P(A) = x + 0.2$, $P(B) = 2x + 0.1$, $P(A \cap B) = x$.

(a) Given that $P(A \cup B) = 0.7$, find the value of x and state the value of x and state the values of $P(A)$ and $P(B)$.

(b) Verify that the events A and B are independent.
The events A and C are mutually exclusive, $P(A \cup B \cup C) = 1$ and $P(B \mid C) = 0.4$.

(c) Find the values of $P(B \cap C)$ and $P(C)$.

(d) Giving a reason, state whether or not the events B and C are independent.

U.L. June 89 Applied Maths (8)

SOLUTION 16

(a) $P(A) = x + 0.2$

$P(B) = 2x + 0.1$

$P(A \cup B) = P(A) + P(B) - P(A \cap B)$

$0.7 = P(A) + P(B) - x$

$0.7 = x + 0.2 + 2x + 0.1 - x$

$0.7 - 0.3 = 2x$

$0.4 = 2x$

$\boxed{x = 0.2}$

$P(A) = x + 0.2 = 0.2 + 0.2 = 0.4$

$P(B) = 2(0.2) + 0.1 = 0.5$

$\boxed{P(A) = 0.4}$ $\boxed{P(B) = 0.5}$

(b) Events A and B are independent if $P(A) \times P(B) = P(A \cap B)$

from (a) $P(A \cap B) = x = 0.2$

$0.4 \times 0.5 = 0.2$

$0.2 = 0.2$

Therefore events A and B are independent

(c) Find $P(B \cap C)$ and $P(C)$

Now $P(B/C) = \dfrac{P(B \cap C)}{P(C)}$

$0.4 = \dfrac{P(B \cap C)}{P(C)}$

$0.4\,P(C) = P(B \cap C)$ ... (1)

Also A and C are mutually exclusive

i.e. $P(A \cap C) = 0.$

Events A, B and C are shown in the following Venn diagram.

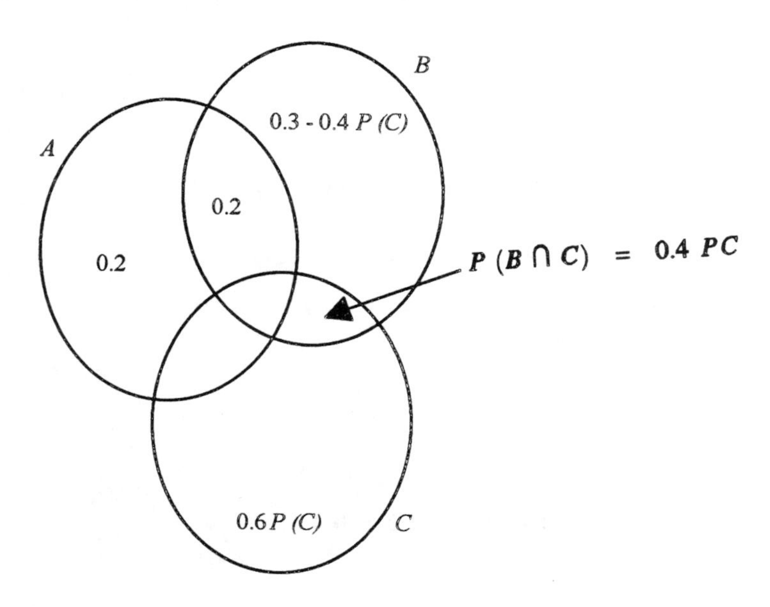

Fig. 10-M/3

Note $P(A \cap B) = 0.4 \quad \therefore P(A \cap \bar{B}) = 0.2$

$P(B) = 0.5 \therefore P(B \cap \bar{A}) = 0.3 - 0.4P(C)$

$P(C \cap \bar{B}) = P(C) - 0.4P(C) = 0.6\,P(C)$

Since $P(A \cup B \cup C) = 1$

So $0.2 + 0.2 + 0.3 - 0.4\,P(C) + 0.4\,P(C) + 0.6\,P(C) = 1$

$$0.7 + 0.6\,P(C) = 1$$

$$0.6\,P(C) = 0.3$$

$$P(C) = \frac{0.3}{0.6}$$

$$P(C) = \frac{1}{2}.$$

Substituting for $P(C)$ in 1

$0.4 \times P(C) = P(B \cap C)$

$$0.4 \times \frac{1}{2} = P(B \cap C)$$

$$0.2 = P(B \cap C)$$

$$\frac{1}{5} = P(B \cap C)$$

(d) If events B and C are independent

$$P(B).P(C) = P(B \cap C)$$

$$\frac{1}{4} \times \frac{1}{2} = \frac{1}{5}$$

$$\frac{1}{8} \neq \frac{1}{5}.$$

So events B and C are not independent.

17. Box A contains 3 blue, 1 yellow and 2 red marbles.
Box B contains 1 blue, 2 yellow and 3 red marbles.
Box C contains 2 yellow and 4 red marbles.
A fair die is thrown. When the score is 1, 2 or 3, box A is chosen. When the score is 4, box B is chosen. When the score is 5 or 6, box C is chosen. A marble is then drawn at random from the chosen box. When this marble is red, it is replaced. When this marble is not red, it is not replaced. A second marble is then drawn at random from the same box.

(a) Find the probability that both marbles drawn will be red.

Given that the two marbles drawn are of the same colour,

(b) find the probability, to 3 decimal places, that the score on the die will be 6.

The experiment is carried out four times.

(c) Find the probability, to 2 decimal places, that on at least one occasion the score on the die is more than 3 and the first marble drawn is not red.

June 1989 Special paper (10)

SOLUTION 17

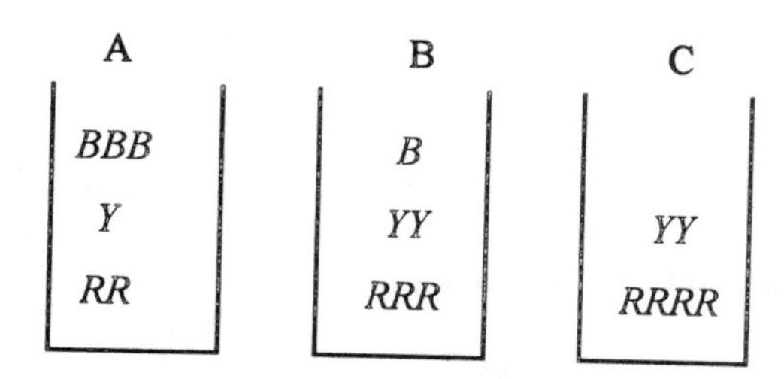

Let A_x be the event of selecting box A

B_x be the event of selecting box B

C_x be the event of selecting box C

B be the event of drawing a blue marble
Y be the event of drawing a yellow marble
R be the event of drawing a red marble
S be the event that two marbles drawn are of the same colour.

(a) P (Both marbles drawn are red)

$$\text{Note } P(A_x) = \frac{3}{6},\ P(B_x) = \frac{1}{6},\ P(C_x) = \frac{2}{6}$$

$$= P\left[(A_x \cap R_1 \cap R_2) \cup (B_x \cap R_1 \cap R_2) \cup (C_x \cap R_1 \cap R_2)\right]$$

$$P(A_x) \times P(R_1) \times P(R_2) + P(B_x) \times P(R_1) \times P(R_2) + P(C_x) \times P(R_1) \times P(R_2)$$

$$= \frac{3}{6} \times \frac{2}{6} \times \frac{2}{6} + \frac{1}{6} \times \frac{3}{6} \times \frac{3}{6} + \frac{2}{6} \times \frac{4}{6} \times \frac{4}{6}$$

$$= \frac{12}{216} + \frac{9}{216} + \frac{32}{216}$$

$$= \frac{53}{216}.$$

(b) P (score of 6 / both marbles are of the same colour) $= P(6/S)$

$$P(6/S) = \frac{P(6 \cap S)}{P(S)}$$

$$P(6 \cap S) = P(6) \cap [P(R_1 \cap R_2) \cup P(Y_1 \cap Y_2)]$$

$$= P(6) \times [P(R_1) \times P(R_2) + P(Y_1) \times P(Y_2)]$$

$$= \frac{1}{6} \times \left(\frac{4}{6} \times \frac{4}{6} + \frac{2}{6} \times \frac{1}{5}\right)$$

$$= \frac{1}{6}\left(\frac{16}{36} + \frac{2}{30}\right) = \frac{23}{270}.$$

$$P(S) = P(\text{Both Red} \cup \text{Both Yellow} \cup \text{Both Blue})$$

$$= P(\text{Both Red}) + P(\text{Both Yellow}) + P(\text{Both Blue})$$

$$P(\text{Both Red}) = \frac{53}{216} \text{ \{from (a)\}}$$

$$P(\text{Both Yellow}) = P(B_x \cap Y_1 \cap Y_1) \cup P(C_x \cap Y_1 \cap Y_2)$$

$$= P(B_x) \times P(Y_1) \times P(Y_2) + P(C_x) \times P(Y_1) \times P(Y_2)$$

$$= \frac{1}{6} \times \frac{2}{6} \times \frac{1}{5} + \frac{2}{6} \times \frac{2}{6} \times \frac{1}{5}$$

$$= \frac{6}{180}$$

$$= \frac{1}{30}.$$

$$P(\text{Both Blue}) = P(A_x \cap B_1 \cap B_2)$$

$$= P(A_x) \times P(B_1) \times P(B_2)$$

$$= \frac{3}{6} \times \frac{3}{6} \times \frac{2}{5} = \frac{1}{10}.$$

$$P(S) = \frac{53}{216} + \frac{1}{30} + \frac{1}{10}$$

$$= \frac{409}{1080}$$

$$\therefore P(6/S) = \frac{P(6 \cap S)}{P(S)}$$

$$= \frac{23/270}{409/1080}$$

$$= \frac{92}{409}$$

$$= 0.225.$$

(c) P (on at least 1 occasion the score > 3 and the first marble is not red)

$$= P(>3 \cap \bar{R})\, P\overline{(>3 \cap \bar{R})^3} + P(>3 \cap \bar{R})^2\, P\overline{(>3 \cap \bar{R})^2} +$$

$$P(>3 \cap \bar{R})^3\, P\overline{(>3 \cap \bar{R})^1} + P(>3 \cap \bar{R})^4$$

$$= 1 - P\overline{(>3 \cap \bar{R})^4}$$

Now $P\overline{(>3 \cap \bar{R})} = 1 - P(>3 \cap \bar{R})$

$P(>3) = P(4)$ is selecting Box B + $P(5 \cup 6)$ selecting Box C.

(i) If one assumes that the two marbles drawn are of the same colour.

So $P(>3 \cap \bar{R}) = P(B_x) \times P(Y_1) + P(C_x) \times P(Y_1)$

$$= \frac{1}{6} \times \frac{2}{6} + \frac{2}{6} \times \frac{2}{6}$$

$$= \frac{2}{36} + \frac{4}{36}$$

$$= \frac{1}{6}$$

$$P\,\overline{(>3 \cap \bar{R})} = 1 - \frac{1}{6}$$

$$= \frac{5}{6}.$$

$\therefore$ P (on at least 1 occasion in 4 trials the score $>$ 3 and the first marble is not Red)

$$= 1 - \left(\frac{5}{6}\right)^4$$

$$= 1 - \frac{625}{1296}$$

$$= \frac{671}{1296} \approx 0.52.$$

OR

(ii) If one assumes that the two marbles drawn are not of the same colour

then $P\,(>3 \cap \bar{R}) = \big(P\,(B_x) \cap P\,(Y \cup B)\big) \cup \big(P\,(C_x) \cap P\,(Y_1)\big)$

$$= \frac{1}{6} \times \frac{3}{6} + \frac{2}{6} \times \frac{2}{6}$$

$$= \frac{7}{36}$$

$$P\,\overline{(>3 \cap \bar{R})} = 1 - \frac{7}{36}$$

$$= \frac{29}{36}$$

P (at least one occasion in 4 trials the score $>$ 3 and the first marble drawn

is not Red) $= 1 - \left(\frac{29}{36}\right)^4$

$$= 1 - 0.421$$

$$= 0.58.$$

18. Find the number of ways a committee of 4 people can be chosen from a group of 5 men and 7 women when it contains
   (a) only people of the same sex,
   (b) people of both sexes and there are at least as many women as men.

June 1994 U.L. P1 (4)

SOLUTION 18

(a) If there are people of the same sex, the number of ways of selecting two men

are, $^5C_4 = \dfrac{5!}{1!\ 4!} = 5$

The number of ways of selecting two women are,

$$^7C_4 = \frac{7!}{3!\ 4!} = \frac{5 \times 6 \times 7}{1 \times 2 \times 3} = 35$$

$5 + 35 = \boxed{40}$ the total number of ways

(b) If there are people of both sexes and there are at least as many women as men then the committee could contain, 2 women and 2 men or 3 women and one man.

The number of ways of selecting 2 women and 2 men is

$= {}^7C_2 \times {}^5C_2$

$$= \frac{7!}{(7-2)!\ 2!} \times \frac{5!}{(5-2)!\ 2!}$$

$= 21 \times 10 = 210$ ways.

The number of ways of selecting 3 women and 1 man is

$= {}^7C_3 \times {}^5C_1$

$$= \frac{7!}{(7-3)!\ 3!} \times \frac{5!}{(5-1)!\ 1!} = 35 \times 5 = 175 \text{ ways}$$

The number of ways of selecting people of both sexes and at least as many women $= 210 + 175$

$= \boxed{385 \text{ ways}}$

19. A taxi, which carries at most four passengers on any journey, makes two journeys in transporting six people from their home to the station. Find the number of different ways in which people for the first journey may be selected.

Jan. 90 U.L. P1 (3)

SOLUTION 19

Take 2 or 3 or 4 passengers

$$^6C_2 = \frac{6!}{4!\ 2!} = 15$$

$$^6C_3 = \frac{6!}{3!\ 3!} = \frac{4 \times 5 \times 6}{1 \times 2 \times 3} = 20$$

$$^6C_4 = \frac{6!}{4!\ 2!} = 15$$

$$15 + 20 + 15 = \boxed{50}$$

20. Find the number of different arrangements that can be made using all eight letters of the word *ROTATION*.
Find the number of these arrangements in which the letters *T* are not consecutive.

June 1989 U.L. P1 (3)

SOLUTION 20

ROTATION. There are eight letters, double 0, and double *T*.

$$\frac{8!}{2!\ 2!} = 10080$$

There are two '*T*'. There are two '*O*'
Rearranging (*TT*) *ROAION*

$$\frac{7!}{2} = 25200$$

(*TT*) ... *TT* are consecutive

The number of ways *TT* not consecutive is
Total number of ways - Number of ways with *TT* consecutive.

$$10080 - 2520 = \boxed{7560}$$

21. Calculate the number of distinct nine letter arrangements that can be made with the letters of the word

$$A \quad R \quad R \quad O \quad W \quad R \quad O \quad O \quad T.$$

Calculate also the number of these arrangements in which A and T are at the two ends with all the R's together and with all the O's together.

Jan. 1989 U.L. P2 (3)

SOLUTION 21

ARROWROOT

There are 9 letters. There are 3 'R'. There are 3 'O'.

$$\text{Number of arrangements} = \frac{9!}{3!\ 3!} = \boxed{10080}$$

Without A and T and the R and O together

there are three terms to permutate with A and T at the ends.

$A(RRR)\ (OOO)\ WT$ and $T(RRR)(OOO)\ WA$

The number of ways to permulate the middle three terms of

$T(RRR)(OOO)\ WA = 3! = 6$ ways

The number of ways to arrange the middle three terms of

$A(RRR)(OOO)\ WA = 3! = 6$ ways

there are therefore $\boxed{12}$ arrangements.

10. COMBINATIONS. PERMUTATIONS. PROBABILITIES

INDEX